# Advanced Automation
# for Comprehensible Causal Explanations
# of Reinforcement Learning Agents

Rudy Milani

# Advanced Automation for Comprehensible Causal Explanations of Reinforcement Learning Agents

 Springer Vieweg

Rudy Milani
Institute for Theoretic Computer Science,
Mathematics and Operations Research
University of the Bundeswehr Munich
Neubiberg, Germany

ISBN 978-3-658-50494-6     ISBN 978-3-658-50495-3   (eBook)
https://doi.org/10.1007/978-3-658-50495-3

This work is funded by dtec.bw—Digitalization and Technology Research Center of the Bundeswehr project RISK.twin. dtec.bw is funded by the European Union—NextGenerationEU.



This Springer Vieweg imprint is published by the registered company Springer Fachmedien Wiesbaden GmbH, part of Springer Nature.
The registered company address is: Abraham-Lincoln-Str. 46, 65189 Wiesbaden, Germany

# Acknowledgments

First and foremost, I would like to express my deepest gratitude to my advisor, Prof. Maximilian Moll, for his invaluable guidance, continuous support, and insightful feedback throughout the course of this research. His expertise and encouragement have been instrumental in shaping both this thesis and my academic journey. I am truly grateful for his patience and unwavering belief in my work.

I would also like to extend my sincere thanks to Prof. Renato De Leone for his constructive input and for serving as an academic father during all these years. His contributions and perspective have been essential in enhancing the quality of this research as well as helping me shaping my academic career. Without his support, I would not have entered this path and finally accomplished my Ph.D.

A very special thanks goes to Prof. Stefan Pickl, the leader of COMTESSA. I will always be grateful for his trust in accepting me in his research group and welcoming me into the Universität der Bundeswehr München. His leadership, vision, and encouragement have been key to the success of this dissertation, and his support has motivated me throughout the challenging moments of this journey.

Furthermore, I would like to thank my Ph.D. commission: Prof. Dr.-Ing. Markus Siegle, Prof. Dr. Michaela Geierhos, Prof. Dr. Gunnar Teege and Prof. Dr. Arno Wacker.

Special thanks to my colleagues and friends from the COMTESSA group. The vibrant discussions and collaboration with all of you have inspired me in many ways and provided a stimulating academic environment.

I am deeply grateful to my family, especially my parents, my brother, and my grandma, for their unconditional love, support, and encouragement throughout this journey. Your belief in me has been my source of strength and motivation.

Lastly, I could not have completed this dissertation without the support of my friends. The ones that I met far away from home, and the ones that were next to my door, the ones that I met during these university years and the ones that grew up with me. I wish to thank you because each of you has left something in my heart.

Thank you all for being part of this incredible journey.

# Abstract

In recent years, the rise in applications of Reinforcement Learning has been fueled by the availability of powerful computational resources and advanced methods, leading to breakthroughs in areas such as autonomous driving, medicine, and finance. However, these advancements often come at the cost of transparency, leading to diminished trust from human users. This thesis addresses this issue by proposing Auto-BENEDICT, a novel methodology designed to automatically generate causal explanations for the actions of model-free Reinforcement Learning agents. The explanations provided answer to both "Why" and "Why not" questions, increasing the comprehensibility of agent decisions. Auto-BENEDICT integrates Bayesian Networks for causal reasoning, and Recurrent Neural Networks for forecasting the distal information, recognized through the adoption of novel importance metrics. Improvements are brought to each of these individual components in order to enhance both computational efficiency and human understanding. The methodology is validated through computational experiments, where the predictive accuracy and scope of the causal models are improved. The human validation study further demonstrates that the explanations generated are statistically superior to traditional methods in terms of comprehensibility and trustworthiness. This research advances the field of Explainable Reinforcement Learning by offering a fully automated process that balances the need for both accurate predictions and human-friendly explanations, thereby improving user trust and satisfaction with Reinforcement Learning agents.

# Contents


**1  Introduction** .................................................. 1
  1.1  Central Research Formulation ............................. 3
  1.2  Contributions and Outline ................................ 5

**2  An Introduction to Reinforcement Learning** .................. 9
  2.1  Reinforcement Learning and Markov Decision Processes ...... 9
  2.2  Value Functions and Bellman Optimality Equation ........... 12
  2.3  Tabular Reinforcement Learning ............................ 15
      2.3.1  Q-learning ......................................... 16
      2.3.2  SARSA .............................................. 17
  2.4  Deep Reinforcement Learning ............................... 19
      2.4.1  Deep Q-Network ..................................... 19
      2.4.2  REINFORCE .......................................... 21

**3  An Overview of Explainable Reinforcement Learning** .......... 25
  3.1  Explainability and Interpretability: The Definitions in XAI ..... 25
  3.2  The Novel Classification Scheme ........................... 27
  3.3  Literature Review of XRL .................................. 28
      3.3.1  Relational and Symbolic RL ......................... 30
      3.3.2  Reward Decomposition ............................... 32
      3.3.3  Hierarchical and Goal-Based Methods ................ 33
      3.3.4  Consequence Based Methods .......................... 34
      3.3.5  Causal Models ...................................... 36
      3.3.6  Human Robot Collaboration .......................... 38
      3.3.7  Visual Based Methods ............................... 39

3.3.8   Policy Simplification .................................   42
3.4   Discussion and Detailed Formulation of Research Questions  ...   45

**4   A Bayesian Network and RNN Methodology for Causal-Distal XRL** .................................................................   51
4.1   Preliminaries ..........................................   52
4.1.1   Reasoning and Causality ...........................   52
4.1.2   Bayesian Networks ................................   58
4.1.3   Learning Causal Structures from Data .................   64
4.2   Specific Literature: An Action Influence Model Approach .....   77
4.2.1   Mathematical Formulation and Practical Examples .....   78
4.2.2   Limitation of the Reviewed Methodology ..............   85
4.3   BENEDICT: An Automatic Distal XRL Approach ............   86
4.3.1   Learning the Causal Structure ......................   88
4.3.2   Learning the BN Probabilities ......................   93
4.3.3   Learning the Distal Information .....................   94
4.3.4   Creating the Explanations ..........................   99
4.4   Computational Experiments ...........................   103
4.4.1   Taxi Environment ................................   104
4.4.2   Cartpole Environment .............................   107
4.4.3   Mountain Car Environment ........................   111
4.4.4   Discussion of the Results ..........................   115
4.5   Human Validation Study ...............................   117
4.5.1   Experimental Settings and Statistical Methodologies ....   118
4.5.2   Statistical Results for "Why" Questions ..............   120
4.5.3   Statistical Results for "Why not" Questions ...........   125
4.5.4   Discussion of the Human Study Results ..............   129
4.6   Answering Research Question 1 .........................   130

**5   Recognition of Important States** ...............................   133
5.1   Preliminaries .........................................   134
5.1.1   Multigraphs and Multilayer Networks .................   135
5.1.2   Centrality Measures .............................   136
5.2   Specific Literature: Importance Metrics in XRL ..............   145
5.2.1   Q-value-based Metrics ............................   145
5.2.2   Graph-based Metrics ..............................   148
5.2.3   Limitations of Existing Measures ...................   150
5.3   Improving Q-value-based Metrics .........................   151
5.3.1   A Novel Metric: The Difference Measure ............   151
5.3.2   IQVA: The Iterated Q-Value Approach ...............   153

5.3.3 Computational Experiments ........................... 155
5.3.4 Discussion ........................................ 162
5.4 Extending the Graph-based Metrics to Multigraphs
and Multiplexes ........................................ 163
5.4.1 The Reconstructed Transition Multigraph and its
Centrality Measures ............................... 163
5.4.2 The Reconstructed Transition Multiplex and its
Centrality Measures ............................... 179
5.4.3 Computational Experiments ........................ 188
5.4.4 Discussion ....................................... 197
5.5 Comparison of Metrics ................................. 198
5.6 Answering Research Question 2 ......................... 201

6 Auto-BENEDICT: Combining Importance Metrics
and BENEDICT for an Automatic Explanation Generator ........ 203
6.1 Coupling Importance Metrics and Explanation Generator:
the Auto-BENEDICT Approach ........................... 204
6.2 Computational Experiments ............................. 208
6.3 Discussion ............................................ 217
6.4 Answering Research Question 3 ......................... 218

7 Explainability for Control of Digital Twins: An Application
of Auto-BENEDICT ......................................... 221
7.1 Digital Twins ......................................... 222
7.2 A Case Study of Digital Twin: The Robotic Arm
Environment ........................................... 224
7.3 Results in the Robotic Arm Environment .................. 225
7.4 Discussion ............................................ 239

8 Conclusion and Future Work ................................ 241
8.1 Contributions ......................................... 241
8.2 Future Works ......................................... 243

Bibliography ................................................ 245

# List of Figures


Fig. 1.1     Structure of the thesis with corresponding topics and research questions addressed for each chapter ......... 6

Fig. 2.1     Schematic representation of the interaction between agent and environment in RL [232] ............. 11

Fig. 2.2     Taxonomy of RL algorithms .......................... 15

Fig. 3.1     Bar plot of the number of articles found using the key word "Explainable Reinforcement Learning" in multiple search engines ...................................... 29

Fig. 3.2     Venn diagrams representing the classification groups shown in the literature review. It is included in dashed line the group of Importance Metrics, due to its relevance in Chap. 5 (Modified from: Advances in explainable reinforcement learning: An intelligent transportation systems perspective, Edition by Rudy Milani, Maximilian Moll, and Stefan Pickl, Copyright ©2023 by Imprint. Reproduced by permission of Taylor & Francis Group) ........................... 46

Fig. 4.1     Example of Causation-Correlation relationship ............ 53

Fig. 4.2     Examples of associated diagrams to the SEM in Equation (4.1). The exogenous variables are recognized by the dashed lines .......................... 54

Fig. 4.3     Diagram associated to the SEM in Equation (4.5) .......... 56

Fig. 4.4     Diagram associated to the SCM in Equation (4.7) .......... 57

Fig. 4.5     Example of DAG for a BN ............................. 59

Fig. 4.6     Action influence graph derived from the Taxi
             environment .......................................................   79
Fig. 4.7     Rendering of the Taxi environment [61] ....................   80
Fig. 4.8     Pipeline of BENEDICT ......................................   87
Fig. 4.9     DAG obtained from NOTEARS-MDP with $\varepsilon = 0.5$ ........   90
Fig. 4.10    DAG obtained from classical algorithms: (a) NOTEARS
             with $\varepsilon = 0.5$, (b) DYNOTEARS with $\varepsilon = 0.01$, (c)
             DYNOTEARS with $\varepsilon = 0.1$, (d) DirectLiNGAM
             with actual features, (e) DirectLiNGAM with actual
             and future features, (f) VARLiNGAM with $\varepsilon = 0.1$ ........   91
Fig. 4.11    Taxi environment with central bottleneck highlighted
             in orange .........................................................   95
Fig. 4.12    Visualization of the different phases of the dataset
             generation process ...............................................   97
Fig. 4.13    Comparison of losses and accuracy of RNNs
             with different cells and features for the Taxi problem .......   98
Fig. 4.14    Procedure for the recognition of the actual
             and counterfactual central variables and their application
             in the identification of the causal chains ...................   100
Fig. 4.15    Training and test losses and accuracy for the Actual
             and Counterfactual RNNs in the Taxi environment .........   105
Fig. 4.16    Structure of the BN for the Cartpole environment. The
             names of the nodes represent the following features:
             "cart pos" is the cart position, "cart vel" is the cart
             velocity, "pole ang" is the pole angle, "pole vel"
             is the pole velocity, "action" is the action, "rew" is
             the reward, "next cart pos" is the next value for the cart
             position, "next cart vel" is the next value for the cart
             velocity, "next pole ang" is the next pole angle,
             "next pole vel" is the next value for the pole velocity ......   108
Fig. 4.17    Training and test losses for the Actual and Counterfactual
             RNNs in Cartpole environment ...........................   109
Fig. 4.18    Structure of the BN for the Mountain Car environment.
             The names of the nodes represent the following
             features: "pos" is the position of the car, "vel"
             is the velocity of the car, "action" is the action,
             "next pos" is the next position of the car, "next vel" is
             the next velocity of the car and "rew" is the reward ........   112

Fig. 4.19   Training and test losses for the Actual and Counterfactual
            RNNs in the Mountain Car environment  . . . . . . . . . . . . . . . . .   113
Fig. 4.20   Statistical methodologies applied for the human
            evaluation study . . . . . . . . . . . . . . . . . . . . . . . . . . . . . . . . . . . . .   119
Fig. 4.21   Box plots of the marks associated with each attribute
            of the explanations for the "Why" question . . . . . . . . . . . . . .   121
Fig. 4.22   Bar chart of the preferences relative to specific qualities
            assigned by the participants to the explanations
            of the "Why" question  . . . . . . . . . . . . . . . . . . . . . . . . . . . . . . .   124
Fig. 4.23   Bar chart representing the results of the true and false
            statements results indicating the characteristics
            of the proposed explanation for the "Why" question . . . . . . . .   124
Fig. 4.24   Box plots of the marks associated with each attribute
            of the explanations for the "Why not" question . . . . . . . . . . .   126
Fig. 4.25   Bar chart of the preferences relative to specific qualities
            assigned by the participants to the explanations
            of the "Why not" question . . . . . . . . . . . . . . . . . . . . . . . . . . . . .   128
Fig. 4.26   Bar chart representing the results of the true and false
            statements results indicating the characteristics
            of the proposed explanation for the "Why not" question  . . . .   128
Fig. 4.27   Radar charts of the mean scores obtained for each
            quality of the explanations for: (a) "Why" and (b) "Why
            not" questions . . . . . . . . . . . . . . . . . . . . . . . . . . . . . . . . . . . . . .   130
Fig. 5.1    Environment used for the test of IQVA algorithms: (a)
            Lava-lake, (b) Key-door, and (c) Taxi  . . . . . . . . . . . . . . . . . .   155
Fig. 5.2    Heatmaps for the Lava-lake environment: measures
            relative to (a) Advising, (b) Action gap, and (c)
            Difference  . . . . . . . . . . . . . . . . . . . . . . . . . . . . . . . . . . . . . . . . .   157
Fig. 5.3    Heatmaps for the Key-door environment (pre-pick
            up key): measures relative to (a) Advising, (b) Action
            gap, and (c) Difference  . . . . . . . . . . . . . . . . . . . . . . . . . . . . . .   158
Fig. 5.4    Heatmaps for the Key-door environment (post-pick
            up key): measures relative to (a) Advising, (b) Action
            gap, and (c) Difference  . . . . . . . . . . . . . . . . . . . . . . . . . . . . . .   159
Fig. 5.5    Heatmaps for the Taxi environment (pre-pick up):
            measures relative to (a) Advising, (b) Action gap,
            and (c) Difference  . . . . . . . . . . . . . . . . . . . . . . . . . . . . . . . . . . .   160

Fig. 5.6 Heatmaps for the Taxi environment (post-pick up): measures relative to (a) Advising, (b) Action gap, and (c) Difference ................................... 161

Fig. 5.7 Example of Reconstructed Transition Multigraph .......... 168

Fig. 5.8 Example of how, by translating all the weights of a constant value, the shortest paths (highlighted) are modified: (a) graph before the translation, (b) graph after translating the weights of 1 ........................ 171

Fig. 5.9 Example of the structural bottleneck edges condition used considered in Theorem 5.5. The bottleneck edge having $r^k_{g-1,g} > 0$ and connecting $s_{g-1}$ to the sub-goal $s_g$ is highlighted ..................................... 172

Fig. 5.10 Example of the structural bottleneck node condition used considered in Theorem 5.6. The edges having positive equal reward $\bar{r} > 0$ and incoming to the bottleneck node $s_g$ are highlighted ..................................... 175

Fig. 5.11 Example of Reconstructed Transition Multiplex ........... 183

Fig. 5.12 Rendering of the custom Taxi environment with traffic ..... 189

Fig. 5.13 Heatmaps of the centrality measures in the simple (unweighted) Reconstructed Transition Graphs and Reconstructed Transition Multigraphs for different positions before picking up the passenger ................ 190

Fig. 5.14 Heatmaps of the centrality measures in the simple (unweighted) Reconstructed Transition Graphs and Reconstructed Transition Multigraphs for different positions after picking up the passenger .................. 192

Fig. 5.15 Heatmaps of the centrality measures in the Reconstructed Transition Multiplex for different positions before picking up the passenger ......................... 194

Fig. 5.16 Heatmaps of the centrality measures in agglomerated states from Fig. 5.15 ................................... 194

Fig. 5.17 Heatmaps of the centrality measures in the Reconstructed Transition Multiplex for different positions after picking up the passenger ....................................... 195

Fig. 5.18 Heatmaps of the centrality measures in agglomerated states from Fig. 5.17 ................................... 195

Fig. 5.19   Heatmaps of the $f$-eigenvector centrality measures
in agglomerated states of the Reconstructed Transition
Multiplex ......................................... 196

Fig. 5.20   Density plot of the eigenvector-related metrics ............ 198

Fig. 5.21   Correlation matrix of the following measures:
betweenness simple graph (BSG), eigenvector
simple graph (ESG), closeness simple graph (CSG),
betweenness multigraph (BMG), eigenvector multigraph
(EMG), closeness multigraph (CMG), betweenness
multiplex (BMP), versatility multiplex (VMP), closeness
multiplex (CMP), f-eigenvector multiplex (f-EMP),
action gap (AG), advising (A), mean (M), difference
(D), iterated action gap (IQVA-AG), iterated advising
(IQVA-A), iterated mean (IQVA-M), iterated difference
(IQVA-D) ......................................... 200

Fig. 6.1   Importance time series of the visited states with marker
peaks associated with the relevant states ©2024 IEEE ...... 204

Fig. 6.2   Pipeline of the *Auto-BENEDICT* approach. ©2024 IEEE .... 208

Fig. 6.3   In the first column, the plots of the average and variance
of the importance metrics are reported. In the second
column, the plots of the differentiation values
with the previous time step can be observed. Lastly,
the histograms with the frequencies of the relevant
states found using $\Delta = 0.02$ are displayed. The metrics
considered are (a) Action Gap, (b) Advising, (c)
Difference, (d) IQVA-AG, (e) IQVA-A, (f) IQVA-D.
©2024 IEEE ......................................... 210

Fig. 6.4   In the first column, the plots of the average and variance
of the importance metrics are reported. In the second
column, the plots of the differentiation values
with the previous time step can be observed. Lastly,
the histograms with the frequencies of the relevant
states found using $\Delta = 0.02$ are displayed. The metrics
considered are (a) Betweenness (Simple Unweighted
Graph), (b) Closeness (Simple Unweighted Graph),
(c) Eigenvector (Simple Unweighted Graph), (d)
Betweenness (Multigraph), (e) Closeness (Multigraph),
(f) f-eigenvector (Multiplex). ©2024 IEEE ................ 211

Fig. 6.5    Box plots of the importance measures associated with risky situations considering the states stored in the analyzed trajectories. ©2024 IEEE ................. 216

Fig. 7.1    Robotic Arm Environment: (a) Physical system, (b) Digital Twin, (c) Grid-world environment (at ground level) ........................ 225

Fig. 7.2    Heatmaps of the centrality measures in the simple Reconstructed Transition Graph for the Robotic Arm environment ...................................... 227

Fig. 7.3    Heatmaps of the centrality measures in the Reconstructed Transition Multigraph for the Robotic Arm environment .... 228

Fig. 7.4    Heatmaps of the betweenness centrality in the Reconstructed Transition Multiplex for the Robotic Arm environment before picking up the package ....................................... 230

Fig. 7.5    Heatmaps of the betweenness centrality in the Reconstructed Transition Multiplex for the Robotic Arm environment after picking up the package ....................................... 231

Fig. 7.6    Heatmaps of the closeness centrality in the Reconstructed Transition Multiplex for the Robotic Arm environment before picking up the package ........................ 232

Fig. 7.7    Heatmaps of the closeness centrality in the Reconstructed Transition Multiplex for the Robotic Arm environment after picking up the package ........................... 233

Fig. 7.8    Heatmaps of the versatility measure in the Reconstructed Transition Multiplex for the Robotic Arm environment before picking up the package ........................ 234

Fig. 7.9    Heatmaps of the versatility measure in the Reconstructed Transition Multiplex for the Robotic Arm environment after picking up the package ........................... 235

Fig. 7.10    Heatmaps of the $f$-eigenvector centrality in the Reconstructed Transition Multiplex for the Robotic Arm environment ........................ 236

Fig. 7.11    DAG obtained from NOTEARS-MDP with $\varepsilon = 0.5$ for the Robotic Arm environment ........................ 237

# List of Tables


Table 1.1 List of own publications on which this thesis is built ....... 5

Table 3.1 Selection of the principal papers for the reviewed literature in the field of XRL following the proposed classification rules. The symbol $\sim$ indicates that it is not clear if that property holds (Modified from: Advances in explainable reinforcement learning: An intelligent transportation systems perspective, Edition by Rudy Milani, Maximilian Moll, and Stefan Pickl, Copyright ©2023 by Imprint. Reproduced by permission of Taylor & Francis Group) ............... 47

Table 4.1 AUROC and accuracy for the prediction of each future feature in the Taxi environment .......................... 104

Table 4.2 AUROC and accuracy for the prediction of each future feature in the Cartpole environment .................... 109

Table 4.3 AUROC and accuracy for the prediction of each future feature in the Mountain Car environment ................. 112

Table 4.4 Comparison of the accuracy (%) in prediction obtained with different methods: LR is the Linear Regression, DT is the Decision Tree, MLP is the Multi-Layer Perceptron, $DP$ is the Decision Policy Tree, $DP_n$ is the Decision Policy Tree without a fixed depth and BN is the Bayesian Network ............................. 116

Table 4.5    Table of the $F$-values and $p$-values obtained
            in the Kruskal-Wallis test for the "Why" question
            explanations ........................................ 122
Table 4.6    Table of the $p$-values obtained from Dunn's test
            for the comparison of the "Why" question explanations.
            The names of the respective groups are shortened
            as follows: N = No Explanation, R = Khan et al. [124],
            C = Madumal et al. [150], B = Milani et al. [160] ........ 123
Table 4.7    Table of the $p$-values obtained from the Kruskal-Wallis
            test for the comparison of the "Why no" question
            explanations ........................................ 127
Table 4.8    Table of the $p$-values obtained from Dunn's test
            for the comparison of the "Why not" question
            explanations. The names of the respective groups
            are shortened as follows: N = No Explanation, C =
            Madumal et al. [149], B = Milani et al. [160] ............ 127
Table 5.1    Q-value-based methods using the formula
            $\max_{a \in \mathbb{A}} Q(s, a) - \xi(s)$, where $\xi(s)$ is
            the interchangeable second term ...................... 148
Table 6.1    List of relevant states detected after the analysis of 1000
            trajectories considering twelve importance metrics.
            For each measure, the number of times that the specified
            state has been individuated as relevant and the particular
            state instantiations (in order, destination, passenger
            position, taxi column and taxi row) are reported. ©2024
            IEEE ................................................. 213

# List of Algorithms


Algorithm 2.1    Q-learning with linear decay $\varepsilon$−greedy method
[232] ...........................................    17
Algorithm 2.2    SARSA with linear decay $\varepsilon$−greedy method [232] ....    18
Algorithm 2.3    DQN with experience replay [168] .................    21
Algorithm 2.4    REINFORCE algorithm [264] .......................    23
Algorithm 4.1    ICA-LiNGAM algorithm [216] .....................    66
Algorithm 4.2    DirectLiNGAM algorithm [217] ...................    69
Algorithm 4.3    VARLiNGAM algorithm [112] .....................    71
Algorithm 4.4    NOTEARS algorithm [273] .........................    74
Algorithm 4.5    DYNOTEARS algorithm [180] .....................    77
Algorithm 4.6    NOTEARS-MDP algorithm ........................    89
Algorithm 5.1    IQVA [159] .......................................    154
Algorithm 5.2    Creation of Reconstructed Transition Multigraph
during RL acting [161] ............................    166
Algorithm 5.3    Creation of Reconstructed Transition Multiplex
during RL acting [161] ............................    181

# Introduction 1

Recently, the applications of Artificial Intelligence (AI) have been spread into multiple areas of everyday lives [193]. Example of remarkable successes are reported in fields such as transportation [94], medicine [116, 270], finance [40], and advertisement [178]. In these scenarios, the principal methodologies applied are related to Machine Learning (ML) which is a branch of AI that includes algorithms able to complete a task by learning directly from data. ML methods are classically grouped according to the specific learning paradigms in Supervised, Unsupervised, and Reinforcement Learning (RL). Specifically for the aim of this thesis, the focus is centered on RL, so the attention is moved to this field.

To accomplish solutions for real-world problems, more complex and powerful deep models have been developed, leading to a decrease in the transparency of their operations. In this way, the resulting approaches become *black-boxes*, i.e., methodologies where the users are capable only of seeing what input the algorithm is requesting and what is the output produced, without any further information helpful for improving its process readability [203]. Hence, a critical trade-off between the performances of the RL agent and its comprehensibility arises. Multiple definitions of this compromise have been associated with the aforementioned concept: e.g., accuracy-interpretability/explainability [203], performance-readability [65], and accuracy-comprehensibility [76]. Despite the variety of names, the notion that is tackled is always the same: how can be obtained efficient models that can be also trusted by humans?

This issue led to an increase of the importance of research in the direction of eXplainable AI (XAI) [1], and more in detail, eXplainable RL (XRL) [65]. The demand for XRL is justified also by human psychological causes. Indeed, a fundamental component for exploiting these models in real-life circumstances is



R. Milani, *Advanced Automation for Comprehensible Causal Explanations of Reinforcement Learning Agents*,
https://doi.org/10.1007/978-3-658-50495-3_1

1

represented by the trust of the users: if a decision maker considers the prediction unreliable, then it will not be used anymore [197]. Not only trust, but a wide range of qualities have to be tested in order to satisfy the key requirements for complete human acceptance. Among all the characteristics to analyze, the most adopted in the user evaluation studies are transparency, reliability, comprehensibility and understandability. However, these aspects are all interconnected: to improve trust, it is necessary to increase transparency and reliability in the process of deciding the next actions [193]. This latter operation can be supported by the creation of efficient explanations that can be easily interpreted by the users.

Thus, a critical role is played by the explanations that can be produced. Having accurate reasoning for the specific operation chosen by an RL algorithm can make the difference in the final consideration of the human decision-maker [252]. There is indeed evidence where faulty reasoning can lead to epistemic distortion, and faulty final decisions [156]. Therefore, crucial for attaining the best evaluations from the users' perspective is to analyze human reasoning. In this way, the statement created for explaining the RL choices can be modelled in the same manner. When humans have to reason in a real-world scenario, they tend to use mental models which have been constructed through the experience accumulated [115]. Thereby, this model can be identified as the knowledge of a person. Nonetheless, the structural characteristics of the model can vary depending on the particular theory examined. For example, a classical mental model can identify the causal patterns present in a particular problem, in order to generate cause-and-effect relationships. Additionally, causal reasoning is effortlessly performed by humans, testifying how they are inherently predisposed to analyzing situations in a cause-effect form [84].

Hence, a research direction that has been explored in XRL is the adoption of causal models for creating explanations of RL agents' choices [149]. The generation of causal models is indeed a well-known problem that has been widely studied in the last decades [181, 183, 184]. In particular, these methodologies can directly learn the causal relationship from the stored data, reducing at minimum the human intervention. Thereby, the possibility of having subjective biases is avoided [119]. Furthermore, causal models can address not only "Why" questions, but also counterfactuals, i.e., a circumstance that could have happened but did not [164]. With the answers to "Why not" questions, a deeper understanding and a larger overview of the action selection process of the RL algorithm can be accessed. Furthermore, multiple choices can be explored with this particular strategy, leading to a characterization for each actionable operation. This ability is crucial especially if the end goal of the explanations is to convince an user.

As already remarked, it is substantial to check the validity of the generated statements. The mark of approval can only be given by users that interface the proposed explanations in a well-structured experimental environment [103]. Specific attributes have to be evaluated in order to address an explanation as optimal. Therefore, avoiding this final validation phase can invalidate the opportunity to apply the proposed methodology in real-world problems.

Following the path delineated in this introductory part, in the next section, the main research directions are discussed.

## 1.1   Central Research Formulation

The main qualities that are central to the field of XAI and, in particular, XRL, are automation and human approval. Nevertheless, these characteristics are not easy to combine [161]. Especially, if other conditions, such as specific kinds of explanations, are required in the overall final procedure. For example, in the case of the causal models considered in XRL, no studies, so far, have been able to couple together the creation of a completely automatic pipeline and the human validation [161]. When a causal structure is determined directly from data through the employment of state-of-the-art methodologies, the outcome obtained consists of an oversimplified explanation. For example, in the works [99, 250] a parsimony property for the causal graph was considered in order to have a small number of features in the answers generated. On the other hand, particular causal models necessitating human experts to be generated can produce explanations regarded by human users as satisfying [148, 149]. However, the existence of people capable of knowing exactly how a specific problem works is not a trivial condition. Although this is a specified case in XRL, also in other subfields the coupling of both automation and the creation of human-acceptable explanations are not extensively achieved. Given the usefulness of causal explanations in efficiently mimicking human reasoning, the focus is shifted to the techniques specialized in this area.

While there are processes of creating a reliable and efficient causal model that is well-known, they do not specifically address the circumstances concerning the RL general settings. Thus, there is a need for improvement by taking into account precise properties attained in the analyzed problem. In this way, not only the causal structure derived will be more consistent, but more accurate predictions can also be provided. These results can increment the perception of the explanations in the user's mind, thus, improving the general qualities required for approving the outcome generated. Nonetheless, the information concerning the performed predictions should involve also relevant data that can summarize the correct trajectory that is chosen by the RL agent. An example of this kind of information has been used in [148], where the *distal actions* have been defined as actions that have to be enabled by a correct sequence of action or state visitation. This notion can be extended also to a wider concept including also the states and particularly important values. Thereby, the set of the data given to the users is larger and, hence, more complete. Nevertheless, the major issue related to the usage of these pieces of information lies in their discovery. Indeed, they can be easily picked by experts in the field, but since the end goal is to produce an automatic method, it must be found a different solution.

An option to achieve the automatic discovery of distal information can be the application of importance metrics. Lately, there has been a rediscovery of how much these measures can help determine relevant situations during and after the training, leading to significant improvements in the agents' performances. For this reason, a multitude of metrics have been developed to address various scenarios, depending on the needs. In detail, these approaches can identify special states that present some characteristics. Thus, each metric can be linked to the property that has to be considered to directly discriminate the important states required. Despite the variety of measures already existing, there are no approaches that take into consideration the qualities associated with the distal states, i.e., being bottlenecks and fundamental to visit to solve the task. Therefore, an extension to the classical measures has to be made, in order to recognize these special states. Then, it is possible to identify the actions that have been selected in these circumstances as the distal actions, since they are strictly dependent on the same chain of actions or state visitation. By doing so, the full distal information can be derived.

Additionally, these resulting measurements can be employed to gain insights relative to the critical levels of each occasion. Indeed, this situational awareness has been implemented in causal models for XRL [255], leading to favourably enhancing the explanations created. For this reason, the data concerning the risk of taking a wrong action in the actual time step can play an essential role. Coupling this piece of information with the distal information and the next step predictions can show to the users the different facets leading the agent to the decision taken. Furthermore, the methodology applied for creating the answers to "Why" questions can be adapted to dealing with counterfactuals, i.e., "Why not" questions. Providing the opportunity to double-check other actions can reassure the human user of the choice of the RL agent, especially, when a direct comparison between the different outcomes is illustrated.

The issues that have been outlined in this section will be addressed and solved in this thesis. To clarify where each matter is tackled, the remaining part of this introduction traces the structure of this work, with particular attention to the contributions produced for each chapter.

## 1.2　**Contributions and Outline**

This thesis's main goal is to propose a completely automatic methodology for the generation of causal explanations for model-free RL agents. Furthermore, the resulting answers can respond to "Why" and "Why not" questions, allowing the human user to have a wider perspective on the particular action selected by the RL algorithm. Moreover, the output of this pipeline has been human-validated through a user evaluation study. More in detail, the contributions of this work, summarizing the four publications and a submitted article shown in Table 1.1, consist of the following improvements:

- definition of a novel classification scheme for grouping approaches in XRL,
- design of a pipeline for the creation of causal explanations for both "Why" and "Why not" questions, through the adoption of Bayesian Networks generated with an improved statistical method,
- extension of the concept of distal action to distal information, and updated process for its forecasting in actual and counterfactual scenarios,
- verified improvements in both computational efficiency of the predictions and qualities of the explanations,
- extension of the Q-value-based importance measures to an iterated version, and conceptualization of novel multigraph- and multiplex-based measures, with associated deep analysis of theoretical guarantees for their existence and uniqueness,

**Table 1.1**　List of own publications on which this thesis is built

| Ch. | Ref. | Title | Type |
|---|---|---|---|
| 3 | [161] | Advances in Explainable Reinforcement Learning: An Intelligent Transportation Systems Perspective | Book Chapter |
| 4 | [159] | A Bayesian Network Approach to Explainable Reinforcement Learning with Distal Information | Journal Paper |
| 5 | [158] | Detection of Important States through an Iterative Q-value Algorithm for Explainable Reinforcement Learning | Conference Paper |
| 5 | [160] | A Multigraph and Multiplex Approach for the Recognition of Important States in MDPs | Journal Paper (Submitted) |
| 6 | [157] | Towards an Automatic Ensemble Methodology for Explainable Reinforcement Learning | Conference Paper |

- combination of importance metrics and previous causal explanation generator for enhancing the automation of the procedure and the realization of risk-aware explanations.

The contributions above are distributed in multiple chapters, following the structure depicted in Fig. 1.1. In the following, a more detailed description of the different topics involved in each part of this thesis is reported.

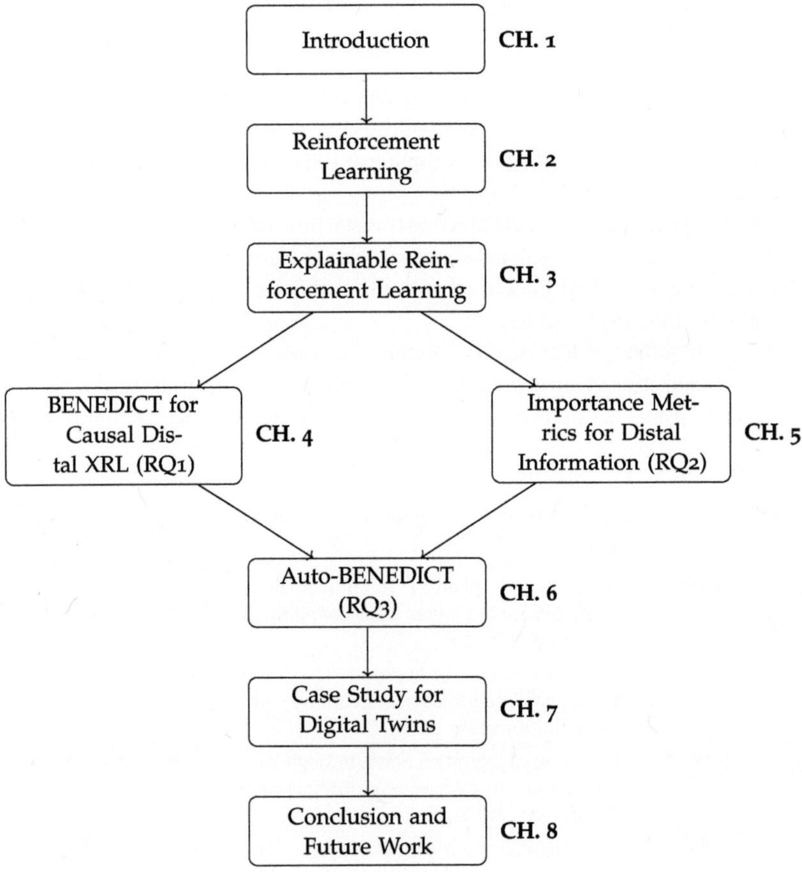


**Fig. 1.1** Structure of the thesis with corresponding topics and research questions addressed for each chapter

The first chapter introduces the motivations and the importance of the problem tackled. Here, the main issues present in the existing methodologies are delineated and the subsequent ideas leading to the contributions are described. Finally, the structure of the thesis is depicted.

In the second chapter, the basic knowledge regarding RL is presented. The description of the theoretical background begins with the classical RL framework represented by the Markov Decision Process. Then, the Bellman Optimality equation is introduced, outlining the most important notions that will be later utilized: the V- and Q-value functions. From these data, it is possible to classify the different methodologies in RL and turn the focus on the model-free approaches. Lastly, both the tabular and deep learning algorithms employed in this work are extensively explained.

After this broad overview of the major RL concepts, the discussion is restricted to the XRL subject and the state-of-the-art methods. In this way, the proposed classification scheme can be shown, comparing it with the attributes considered in other reviews and surveys. By this novel grouping approach, it is possible to spot the main advantages and pitfalls of each class, thus, leading to the recognition of the research gaps. Thereby, the research questions central to this thesis are listed and motivated through specific problems found in the literature.

With the definition of the research questions, the thesis enters the focal part of the methodological discussion with the fourth and fifth chapters. In the former, the problem of automatically creating a causal explanation for RL agents is tackled through the use of the BENEDICT approach: a method combining Bayesian Networks and Recurrent Neural Networks for solving the aforementioned task. In more detail, it is discussed how the answers to the "Why" and "Why not" questions have been enriched with the newly conceptualized distal information. Specifically, the forecasting procedure of these data is improved with special modifications in the preprocessing phase and the selection of the input. Thus, not only the value-based algorithms but also the policy-based methods can be explained by the suggested approach. Then, the construction of the textual statement is declared, showing also examples in multiple environments. Lastly, the results of the evaluation study concerning the comprehensibility of the BENEDICT's outputs are reported.

The fifth chapter is still a methodological chapter, where the main points discussed are the introduction of novel metrics for both the Q-value-based and graph-based measures. In the first case, the classical Q-value-based measures are extended in iterated versions. With this additional property, the states that are frequently visited will attain higher importance scores. Moreover, a new standard metric is derived from the combination of two classical measures: the difference metric. This achievement poses the basis for analyzing the risk of each state. On the other hand, for the graph-based metrics, newly idealized structures for the reconstructed transition graphs are introduced: the Reconstructed Transition Multigraph and the Reconstructed Transition Multiplex. For both these networks, the theoretical assumptions required for ensuring the existence and uniqueness of the associated centrality measures are proved.

The methodologies created in the previous two chapters are then coupled together, obtaining the final completely automatic approach: Auto-BENEDICT. In particular, the sixth chapter discusses how the previously described importance measures can be used for detecting the distal states and for determining the risk levels of the actual instantiation. The risk-aware explanations produced by the Auto-BENEDICT method are finally tested in the Taxi problem.

Subsequently, Auto-BENEDICT is tested in a specific case scenario dealing with a digital twin of a robotic arm in the seventh chapter. In this way, a practical application is also analyzed and the obtained results are discussed.

Lastly, in the eighth and final chapter, all the work done is summarized, with particular highlights on the contributions of the overall research. Additionally, potential future directions of research are listed, proving how the field of XRL presents still multiple grey areas that have to be explored.

# An Introduction to Reinforcement Learning 2

Before turning the attention in the direction of XRL to answer the question of how to automatically generate causal explanations for RL agents, it is necessary to set the background regarding RL. Therefore, in this chapter, the theoretical information fundamental to understanding the classical problems addressed in this thesis is introduced.

The general context in which RL methods are able to learn complex tasks is explained, first identifying the mathematical framework necessary for this end: the Markov Decision Process. From this starting point it is possible to describe the different value functions that will be central to this work, especially for Chap. 5. After this brief introduction to the classical RL setting, the algorithms considered in this thesis are delineated. Specifically, first the tabular methodologies of Q-learning and SARSA are described. Deep learning techniques regarding the approximation of the state-action value function or the policy are also discussed in Deep Q-Network and REINFORCE sections, respectively, Sect. 2.4.1 and 2.4.2.

## 2.1 Reinforcement Learning and Markov Decision Processes

RL is a branch of ML which focuses on solving sequential problems. While for the other two learning paradigms (supervised and unsupervised) the schematic idea of the training process is similar, here a more complex interaction between different entities is involved. Indeed, the RL problem can be framed as the interplay between a decision-maker called *agent* and an *environment*, which represents the rest of the scenario. The task of the agent is to select actions to be performed, and the



R. Milani, *Advanced Automation for Comprehensible Causal Explanations of Reinforcement Learning Agents*,
https://doi.org/10.1007/978-3-658-50495-3_2

environment responds to those choices by presenting new situations. In addition to a new state, the environment also provides a signal that quantifies the quality of an action. This numerical value is usually called reward and provides the basis for learning and improving the behavior of the agent. In this way, the agent tries to maximize the rewards obtained during the episode, i.e., a complete instance of an RL problem.

The learning process outlined above can be formalised by adopting the concept of *Markov Decision Processes* (MDPs). For the aim of this thesis, the description of MDPs is restricted to finite- and discrete-time cases, owing to the characteristics of the practical scenarios treated in this work. Before presenting the technical details of the MDPs, it should be noted that the description of the framework and the notation adopted are based on [231]. Moreover, the latter reference is avoided, reducing citations to other works only to necessary cases.

Formally, at each time step $t = 0, 1, \ldots, T$ the agent gets the actual state $s_t \in \mathbb{S}$ from the environment, and, subsequently, selects an action $a_t \in \mathbb{A}$ depending on it, where $\mathbb{S}$ and $\mathbb{A}$ are the (discrete) state and action spaces, respectively. Consequently, an update of the state and reward is generated. Specifically, the environment presents a new state $s_{t+1} \in \mathbb{S}$ and a numerical reward $r_{t+1} \in \mathbb{R}$ depending on this transition. The classical schematic summary of this interaction is shown in Fig. 2.1. The choice to denote the reward due to $a_t$ as $r_{t+1}$ is meant to emphasize the fact that the reward and the next state are jointly determined. Specifically, the reward is produced by considering the function $r : \mathbb{S} \times \mathbb{A} \times \mathbb{S} \rightarrow \mathbb{R}$ that associates each state, action performed and next state to a compensation, i.e., $r_{t+1} = r(s_t, a_t, s_{t+1})$. Even though this representation is not considering all the general cases since it assumes a deterministic reward given the next state, it is well-posed for the problems that are tackled in this thesis. Additionally, it is also possible to consider the effects of probabilistic dependencies related to state transitions [71]. To identify how the states change, it is fundamental to introduce a formal transition function, $\mathbb{T} : \mathbb{S} \times \mathbb{A} \times \mathbb{S} \rightarrow [0, 1]$. For $\mathbb{T}$ to be a proper transition function, certain properties must hold. The first is the normalisation condition, which states that $\sum_{s_{t+1} \in \mathbb{S}} \mathbb{T}(s_t, a_t, s_{t+1}) = 1$. Nonetheless, $\mathbb{T}$ has to satisfy also the *Markov property*, i.e., for a given history $(s_0, a_0, s_1, \ldots, s_t)$ the transition probability to the next state $s_{t+1}$ only depends on the current state $s_t$ and $a_t$, which is stated (with an abuse of notation, considering $\mathbb{T}$ as a probability) as follows:

$$\mathbb{T}(s_{t+1}|s_0, a_0, s_1, \ldots, s_t, a_t) = \mathbb{T}(s_{t+1}|s_t, a_t) = \mathbb{T}(s_t, a_t, s_{t+1}), \qquad (2.1)$$

where $\mathbb{T}(s_{t+1}|s_t, a_t)$ represents the conditional probability of transitioning to $s_{t+1}$ from $s_t$ performing action $a_t$. An observation can be made regarding the Markov property and the reward function. In fact, by assuming that this condition holds for the transition function, then it is ensured that the generated reward also depends only on the last step information. Moreover, with an abuse of notation, the transition function will consider, on specific occasions, the probability associated with selecting the next state and also the next reward, i.e., $\mathbb{T}(s_{t+1}, r_{t+1}|s_t, a_t) = \mathbb{T}(s_t, a_t, s_{t+1})$. This is allowed because the reward function depends only on the previous state and action, and since the Markov property is satisfied. Nonetheless, the Markov property is also important for guaranteeing the convergence of most algorithms [233]. Finally, a discount factor $\gamma \in [0, 1]$ must be introduced to measure how much the agent value the future rewards compared to immediate rewards. Thereby, the agent's goal is to maximize the return

$$G_t = \sum_{k=0}^{T-t-1} \gamma^k r_{t+k+1},$$

(2.2)

i.e., the total discounted reward obtained from $t$. With this information, the definition of an MDPs can be stated as follows:

$$(\mathbb{S}, \mathbb{A}, r, \mathbb{T}, \gamma),$$

(2.3)

where $\mathbb{S}$ is the state set, $\mathbb{A}$ is the action set, $r$ is the reward function, $\mathbb{T}$ is the transition function and $\gamma$ is the discount factor.

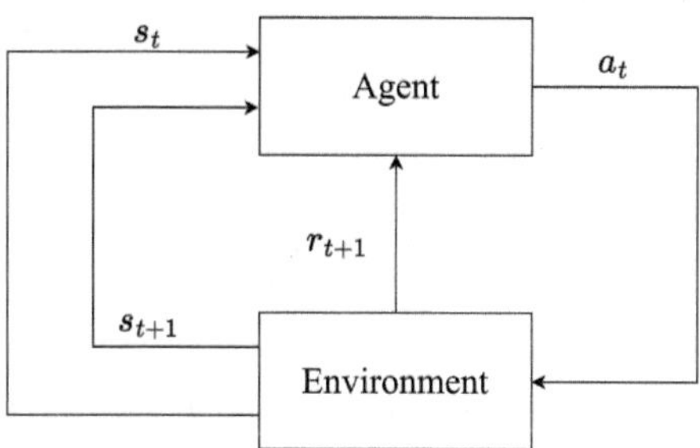


**Fig. 2.1** Schematic representation of the interaction between agent and environment in RL [231]

After presenting the MDP context, the general solution to an RL problem has to be described. After the training of an RL agent, the outcome consists of a *policy* $\pi$, which is a function that maps the states into actions. Thus, by adhering to the *optimal policy* $\pi^*$, the agent can take the best actions in each situation. Mathematically, the policy is defined as either a deterministic function $\pi : \mathbb{S} \rightarrow \mathbb{A}$ or as a probabilistic distribution over actions $\pi : \mathbb{S} \times \mathbb{A} \rightarrow [0, 1]$, where $\sum_{a \in \mathbb{A}} \pi(a|s) = 1$, $\forall s \in \mathbb{S}$. The latter formulation will be adopted in the description of the policy gradient method in Sect. 2.4.2. Among all policies, those that achieve the maximum expected return are identified as *optimal policies*. To obtain the optimal policies, initial policies can be updated according to the experience accumulated during the exploration phase of the environment. However, as reported in the next section, learning the optimal policy is not the only possible choice for solving the RL problem. A common alternative approach is to approximate specific value functions that can summarize the future information, and then select actions according to their output. This topic is central in the subsequent pages.

## 2.2   Value Functions and Bellman Optimality Equation

The value functions involved in RL are associated with the task of estimating the "goodness" of being in a specific state or performing a particular action in that state. However, it is fundamental to precisely address how to identify this quality. Clearly, this notion is strictly related to the reward gained by the agent in future steps. Nevertheless, the selection of actions can generate different results; hence, these value functions must be dependent on the policies chosen by the agents. Therefore, given an MDP, the *state-value function* (or $V$–function) following policy $\pi$ is defined (in the stationary case) as:

$$V_\pi(s) = \mathbb{E}_\pi[G_0|s_0 = s] = \mathbb{E}_\pi\left[ \sum_{k=0}^{T-1} \gamma^k r_{k+1}|s_0 = s \right], \quad \forall s \in \mathbb{S}, \qquad (2.4)$$

where $\mathbb{E}_\pi[\cdot]$ represents the expected value of a random variable, given that the agent will, at each time step, select actions according to the policy $\pi$. Thus, the state-value function associates the expected return when starting at time $t$ in $s$ and, afterwards, following policy $\pi$.

As anticipated at the beginning of this section, a similarly defined value function can be considered, that can also take into account the action performed at $t$. This function is called the *action-value function* (or $Q$–function), and is defined as:

$$Q_\pi(s, a) = \mathbb{E}_\pi[G_t|s_0 = s, a_0 = a] = \mathbb{E}_\pi\left[\sum_{k=0}^{T-t-1} \gamma^k r_{t+k+1}|s_0 = s, a_0 = a\right],$$

$$\forall s \in \mathbb{S}, \ \forall a \in \mathbb{A}.$$

$$(2.5)$$

These value functions have to be estimated from the experience gathered during the agent's training. The simplest way of learning them is to store an average of the actual returns obtained following policy $\pi$ for each state visited. The average converges to the state-value function as the number of visits to each state goes to infinity. Similarly, the action-value function can be approximated by separating the averages for each selected action. Methodologies that apply this idea are called the *Monte Carlo methods*. However, there are also other solutions to this estimation problem that are more practical. An example is provided by the temporal difference (TD) learning which allows direct updates to the value functions based on the estimated TD errors through bootstrapping. Another solution involves the adoption of a parameterized function approximator. RL methods employing both the ideas above are reported in the following sections.

Before focusing on the various algorithms utilized in this thesis, it is helpful to characterize some of their properties: In both the value functions, the formulations can be rearranged in order to consider their recursive expression. For any policy $\pi$ and state $s$, the following series of equalities holds:

$$V_\pi(s) = \mathbb{E}_\pi[G_t|s_t = s] \tag{2.6}$$

$$= \mathbb{E}_\pi\left[\sum_{k=0}^{T-t-1} \gamma^k r_{t+k+1}|s_t = s\right] \tag{2.7}$$

$$= \mathbb{E}_\pi\left[r_{t+1} + \gamma G_{t+1}|s_t = s\right] \tag{2.8}$$

$$= \sum_a \pi(s|a) \sum_{s'} \sum_r \mathbb{T}(s', r|s, a)\left[r_{t+1} + \gamma \mathbb{E}_\pi\left[G_{t+1}|s_{t+1} = s'\right]\right] \tag{2.9}$$

$$= \sum_a \pi(s|a) \sum_{s'} \sum_r \mathbb{T}(s', r|s, a)\left[r_{t+1} + \gamma v_\pi(s')\right], \tag{2.10}$$

where Equations (2.6) and (2.7) are the definitions of the $V-$function. Then, in Equation (2.8) the reward at time $t + 1$ is removed from the summation and the remaining part describes the return at $t + 1$ multiplied by the discount factor. Next, the expectation is expanded in Equation (2.9) and, by recognizing the state-value function evaluated in $s'$, it is possible to deduce the last equality in Equation (2.10). The final result is the well-known *Bellman equation* [26] that defines the relationship between the value of a state and the values of its successor states. By solving this equation, an optimal policy can be computed.

This is guaranteed by the partial ordering of policies defined by the value functions. In other words, a policy $\pi$ that achieves an expected return higher than or equal to another $\pi'$ for all states, i.e., $\pi \geq \pi'$, is equivalent to satisfying $V_\pi(s) \geq V_{\pi'}(s)$, $\forall s \in \mathbb{S}$. Given this property, the value function associated with the optimal policy $\pi^*$ satisfies

$$V_{\pi^*}(s) = \max_\pi V_\pi(s), \quad \forall s \in \mathbb{S}. \tag{2.11}$$

Nonetheless, the same condition holds for the action-value function:

$$Q_{\pi^*}(s, a) = \max_\pi Q_\pi(s, a), \quad \forall s \in \mathbb{S}, \forall a \in \mathbb{A}. \tag{2.12}$$

The optimal action-value function can be rewritten considering that it represents the expected return for taking action $a$ in state $s$, and subsequently, following the optimal policy $\pi^*$. Hence, $Q_{\pi^*}$ can be expressed in terms of $V_{\pi^*}$ as follows:

$$Q_{\pi^*}(s, a) = \mathbb{E}_{\pi^*}\big[r_{t+1} + \gamma V_{\pi^*}(s_{t+1})|s_t = s, a_t = a\big], \forall s \in \mathbb{S}, \forall a \in \mathbb{A}. \tag{2.13}$$

Moreover, if both functions consider optimal policies, then, in general:

$$V_{\pi^*}(s) = \max_{a \in \mathbb{A}} Q_{\pi^*}(s, a), \quad \forall s \in \mathbb{S}. \tag{2.14}$$

From the combination of Equations (2.13) and (2.15), the *Bellman's principle of optimality* is derived as follows:

$$V_{\pi^*}(s) = \max_{a \in \mathbb{A}} \mathbb{E}_{\pi^*}\big[r_{t+1} + \gamma V_{\pi^*}(s_{t+1})|s_t = s, a_t = a\big], \quad \forall s \in \mathbb{S}. \tag{2.15}$$

This relation asserts that the value of a state under an optimal policy must equal the expected return for the best action from that state. Moreover, this equation is fundamental for defining the iterative methodologies addressed in the next section. After defining the multiple value functions and their properties, it is crucial to outline how to solve the MDP for an optimal policy.

## 2.3    Tabular Reinforcement Learning

The algorithms used in RL for solving the MDP for an optimal policy can be categorized as *model-based* and *model-free*. In the former group, the considered models are used to predict the outcomes of the choice of an action in a specific state. In this way, it is possible to plan according to the future step forecasts. The approaches related to the latter class, directly learn without the planning step. Thus, they are not relying on any predictive models [262]. A further division in the model-free algorithms can be identified when the methodologies are based principally on the estimation of the value functions (*value-based*) or on the proper learning of the optimal policy (*policy-based*). Examples of both approaches are reported in the following sections, while the general taxonomy is summarized in Fig. 2.2.

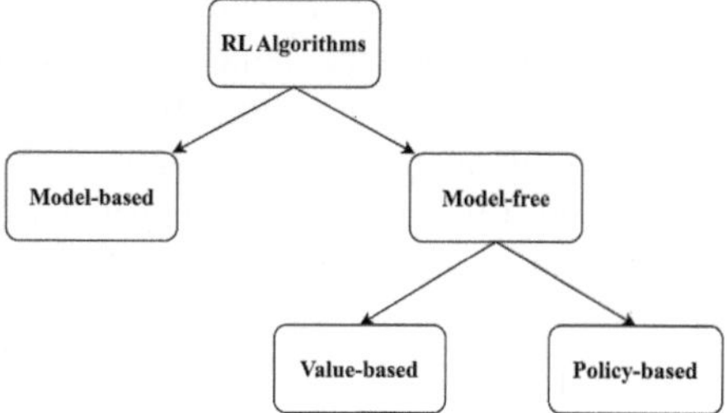


**Fig. 2.2** Taxonomy of RL algorithms

However, before introducing the first algorithms, it is indispensable to address a core problem in RL training: the *exploration-exploitation* trade-off. This problem can be described as follows: while, at the beginning of the learning process, the agent needs to explore the environment by randomly selecting some actions since no reward has been observed yet, in the following episodes, it has to balance between exploring new strategic paths and reproducing the same successful actions as done in the previous episodes. Obviously, it is not guaranteed that, after obtaining the first solution, it is optimal. Therefore, in order to check the quality of the solution found, the agent has to trade off between exploring new situations and exploiting the already known trajectories.

A classical approach to this dilemma is to apply the $\varepsilon-$greedy method: with a probability $\varepsilon \in [0, 1]$ a random action is selected and in the remaining scenarios, the agent greedily chooses the action that corresponds to the highest action-value-function value at that state. The value $\varepsilon$ is usually initialized as a large value and it decays over time in order to increase the exploration at the initial stages of the training and reduce it towards the end of the process.

As previously mentioned, RL adopts principally different techniques to approximate a function from the data generated by the interaction of the agent and the environment. A simple solution can be characterized by the use of tabular methodologies, where through the use of a matrix it is possible to estimate the values of the action-value function. In the following, two of the first methodologies considering this idea are discussed.

### 2.3.1 Q-learning

Despite being one of the earliest algorithms in RL, *Q-learning* [258] still remains one of the most common methodologies considered nowadays. This is due to its simple but effective structure that does not rely on complex function approximations. On the contrary, this methodology leverages the Bellman equation and bootstrapping.

Specifically, as the name suggests, this algorithm learns the action-value function and derives a policy directly by taking the action maximizing that function. Q-learning uses the TD concept: it updates the approximated action-value function through a combination of the observed reward and the estimation of the next state's value. This bootstrapping with the temporal difference process can generate updates that are based on the estimated TD error. Therefore, the updating phase must not wait until the episode is finished as in the Monte Carlo approach. Moreover, since Q-learning is an off-policy methodology, meaning that the approximation does not depend on the policy that the agent is using to sample from the environment, it can learn the optimal policy even if the behavior policy is exploratory. Instead, for on-policy algorithms, the updates are dependent on the adopted policy. In the following section, an example of this type of algorithm is described.

The general procedure of the algorithm consists of the following steps. At the beginning of the learning process, a matrix defined as *Q-table* ($Q(s_t, a_t)$), with dimensions $|\mathbb{S}| \times |\mathbb{A}|$, is initialized as a null matrix. After selecting action $a_t$ at state $s_t$, the Q-table is updated using the following rule:

$$Q(s_t, a_t) \leftarrow Q(s_t, a_t) + \alpha \left[ r_{t+1} + \gamma \max_{a \in \mathbb{A}} Q(s_{t+1}, a) - Q(s_t, a_t) \right], \qquad (2.16)$$

where $\alpha \in (0, 1]$ is the learning rate, i.e., a measure to define the weighting between the old value and the updates. In particular, the update rule is derived considering the idea of Bellman's principle of optimality in Equation (2.15). The full algorithm is reported including the linear decay $\varepsilon-$greedy method for the exploration in Algorithm 2.1.

---

**Algorithm 2.1** Q-learning with linear decay $\varepsilon-$greedy method [231]

---

**Require:** Initial matrix $Q(s, a) \in \mathbb{R}^{|\mathbb{S}|\times|\mathbb{A}|}$, $Q(terminal, \cdot) = 0$, Learning rate $\alpha \in (0, 1]$, Number of episodes $N \in \mathbb{N}$, Discount factor $\gamma \in [0, 1]$.

1: **for** $n = 1, \ldots, N$ **do**
2:     Initialize $\varepsilon = \frac{N-n}{N}$
3:     Initialize $s$
4:     **while** $s$ not $terminal$ **do**
5:         Sample action $a$ as

$$a \leftarrow \begin{cases} \arg\max_{\bar{a}\in\mathbb{A}} Q(s, \bar{a})), & \text{with probability } 1 - \varepsilon, \\ \text{random action}, & \text{with probability } \varepsilon, \end{cases}$$

$\qquad\qquad\qquad\qquad\qquad\qquad\qquad\qquad\qquad\qquad\qquad \triangleright \varepsilon-\text{greedy selection}$

6:         Take action $a$ and observe $r$ and $s'$
7:         Update the Q-matrix

$$Q(s, a) \leftarrow Q(s, a) + \alpha\left[r + \gamma \max_{\bar{a}\in\mathbb{A}} Q(s', \bar{a}) - Q(s, a)\right]$$

$\qquad\qquad\qquad\qquad\qquad\qquad\qquad\qquad\qquad\qquad\qquad \triangleright Q-\text{learning updates}$

8:         Update the state $s \leftarrow s'$
9:     **end while**
10: **end for**
11: **return** $Q(\cdot, \cdot)$

---

## 2.3.2  SARSA

Similarly to Q-learning, an on-policy version of the previous algorithm can be derived: the SARSA algorithm [204]. Here, the major difference with the aforementioned method is in the update rule. While Q-learning updates the Q-values accordingly to the optimal values, in SARSA the updating increments are dependent on the behavior policy adopted. Therefore, the policy used for the choice of the next action is the one that is changing. The learning process consists of improving the selection process of the behavior policy step-by-step. Thus, the resulting update rule is the following:

$$Q(s_t, a_t) \leftarrow Q(s_t, a_t) + \alpha\left[r_{t+1} + \gamma Q(s_{t+1}, a_{t+1}) - Q(s_t, a_t)\right], \qquad (2.17)$$

where $a_{t+1}$ is the action chosen following the behavior policy. As it can be noticed from the updating formula in Equation (2.17), at each iteration a quintuple of events $(s_t, a_t, r_{t+1}, s_{t+1}, a_{t+1})$ is required. Thus, Sutton decided to rename this approach (which was initially known as *Modified Connectionist Q-Learning* [205]) as SARSA because of this quintuple [231]. All the other steps of this methodology are the same as in Q-learning. The SARSA algorithm including the linear decay $\varepsilon-$greedy method for the exploration is presented in Algorithm 2.2.

---

**Algorithm 2.2** SARSA with linear decay $\varepsilon-$greedy method [231]

---

**Require:** Initial matrix $Q(s, a) \in \mathbb{R}^{|\mathbb{S}| \times |\mathbb{A}|}$, $Q(terminal, \cdot) = 0$, Learning rate $\alpha \in (0, 1]$, Number of episodes $N \in \mathbb{N}$, Discount factor $\gamma \in [0, 1]$.

1: **for** $n = 1, \ldots, N$ **do**
2:     Initialize $\varepsilon = \frac{N-n}{N}$
3:     Initialize $s$
4:     Sample action $a$ as

$$a \leftarrow \begin{cases} \arg\max_{\bar{a} \in \mathbb{A}} Q(s, \bar{a})), & \text{with probability } 1 - \varepsilon, \\ \text{random action}, & \text{with probability } \varepsilon, \end{cases}$$

$\qquad\qquad\qquad\qquad\qquad\qquad\qquad\qquad\qquad\qquad\triangleright \varepsilon-$greedy selection

5:     **while** $s$ not *terminal* **do**
6:         Take action $a$ and observe $r$ and $s'$
7:         Sample next action $a'$ as

$$a' \leftarrow \begin{cases} \arg\max_{\bar{a} \in \mathbb{A}} Q(s', \bar{a})), & \text{with probability } 1 - \varepsilon, \\ \text{random action}, & \text{with probability } \varepsilon, \end{cases}$$

$\qquad\qquad\qquad\qquad\qquad\qquad\qquad\qquad\qquad\qquad\triangleright \varepsilon-$greedy selection

8:         Update the Q-matrix

$$Q(s, a) \leftarrow Q(s, a) + \alpha \left[ r + \gamma Q(s', a') - Q(s, a) \right]$$

$\qquad\qquad\qquad\qquad\qquad\qquad\qquad\qquad\qquad\qquad\triangleright \text{SARSA updates}$

9:         Update the state $s \leftarrow s'$
10:         Update the action $a \leftarrow a'$
11:     **end while**
12: **end for**
13: **return** $Q(\cdot, \cdot)$

---

The benefit of using these tabular methods for simple problems is that they can successfully solve the tasks in a reasonable amount of training iterations. However, they suffer from the *curse of dimensionality* [25]: since the table initialized at the beginning of these algorithms scales with $|\mathbb{S}| \times |\mathbb{A}|$, it becomes intractable in larger environments. Moreover, similar issues could arise from continuous problems, where a discretization has to be performed before applying the aforementioned approaches. Hence, this condition hinders the applicability of tabular methodologies for too large state and action spaces. A different way of approximating the value functions or policies will be discussed in the next section.

## 2.4   Deep Reinforcement Learning

In recent years, a high interest in Deep Learning sparked due to the extraordinary applicability of these models [273]. In particular, the class of Artificial Neural Networks (ANNs) has been able to successfully complete a wide spectrum of complex problems, ranging from speech and image recognition to forecasting time series [18]. Between those, the ability of the ANNs to be an excellent function approximator stands out. In fact, when a sigmoidal function is utilized as an activation function, the ANN is a universal function approximator [52, 104]. Thus, such ANNs can approximate any Lebesgue integrable function, assuming that the depth of the network is sufficient. From this aspect, derived an increasing interest in Deep Neural Networks (DNNs), and, consequently, Deep Reinforcement Learning (DRL) when applied to the previously described estimation problems in RL.

In fact, this particular approach can be applied in model-free RL for two specific situations: to approximate the value function (Sect. 2.4.1) or directly the policy (Sect. 2.4.2). In the following sections, an example for each of these cases is reported.

### 2.4.1   Deep Q-Network

As already stated, the tabular algorithms suffer from the curse of dimensionality, so, it is beneficial to determine a more suitable approximator for learning the value function. In this scenario, DNNs can represent a solution given that they require less memory compared to Q-learning and SARSA. Nevertheless, the usage of DNNs for estimating the value function in temporal-difference learning produces an unstable behavior [240]. This problem is, together with bootstrapping and off-policy learning, one of the *deadly triad* (the term was coined by Sutton and Barto [231]): properties that, if combined, can significantly disrupt the stable and efficient learning of optimal policies. Specifically, a suitable training sample and the correct action-value function estimation are necessary to avoid this issue.

The solution, that led to the nowadays applied algorithm, was found by Mnih et al. [166]. Here, the authors included a memory in the agent to store the information related to the states, actions and reward sequences that can be later used as a replay buffer. Through this replay buffer, random samples of mini-batches of experience are drawn and used for the training of the ANN. With this strategy, the temporal correlations within the data are removed. This peculiar procedure is possible only in Q-learning, due to its property of being an off-policy method.

The full algorithm is based on identifying the approximate action-value function $Q(s, a; \theta_i)$, where $\theta_i$ are the weights of the ANN (called Q-network) at iteration $i$. The loss function used for the updating of the weights, derived from Q-learning, is defined as follows:

$$L_i(\theta_i) = \mathbb{E}_{s,a,r,s'}\left[ (y_i - Q(s, a, ; \theta_i))^2 \right], \tag{2.18}$$

where $y_i = r + \gamma \max_{\bar{a}\in\mathbb{A}} Q(s', \bar{a}, \bar{\theta}_i)$ is the target at iteration $i$ and $\bar{\theta}_i$ are the network parameters used to compute the target at iteration $i$. These target weights are only updated with the Q-network once every $C$ steps and remain fixed between individual updates. With this choice of the target, it is shown that the training is stabilized [167]. Then, by differentiating the loss with respect to the weights, we obtain:

$$\nabla_{\theta_i} L_i(\theta_i) = \mathbb{E}_{s,a,r,s'}\left[ (r + \gamma \max_{\bar{a}\in\mathbb{A}} Q(s', \bar{a}; \bar{\theta}_i) - Q(s, a; \theta_i))\nabla_{\theta_i} Q(s, a; \theta_i) \right], \tag{2.19}$$

which is used in the stochastic gradient descent step for optimizing the Q-network parameters. Given the application of DNNs in approximating the action-value function, this algorithm, summarized in Algorithm 2.3, is referred to as *Deep-Q-Network* (DQN).

Despite performing at super-human levels in a wide range of Atari games [166], this methodology presents some limitations. In detail, it is possible to get stuck in poor local minima or even diverge [257]. Another major challenge of DQN is to learn temporally long strategies like in Montezuma's revenge game. Furthermore, it is unfeasible to generate continuous policies. However, for this latter remark, a solution can be given by considering the *policy gradient* methods. An example of an algorithm belonging to this class is reported in the following section.

---

**Algorithm 2.3** DQN with experience replay [167]

---

**Require:** Action-value function with weights $\theta$, Initialization $Q(s, a; \theta) \in \mathbb{R}, \forall s \in \mathbb{S}, \forall a \in \mathbb{A}$, Target action-value function with weights $\bar{\theta}$, $Q(s, a; \bar{\theta}) \in \mathbb{R}, \forall s \in \mathbb{S}, \forall a \in \mathbb{A}$, Replay Memory $D$, Learning rate $\alpha \in (0, 1]$, Number of episodes $N \in \mathbb{N}$, Discount factor $\gamma \in [0, 1]$.

1: **for** $n = 1, \ldots, N$ **do**
2:     Initialize $\varepsilon = \frac{N-n}{N}$
3:     Initialize $s$
4:     **while** $s$ not $terminal$ **do**
5:         Sample action $a$ as

$$a \leftarrow \begin{cases} \arg\max_{\bar{a}\in\mathbb{A}} Q(s, \bar{a})), & \text{with probability } 1 - \varepsilon, \\ \text{random action}, & \text{with probability } \varepsilon, \end{cases}$$

$$\rhd \, \varepsilon\text{-greedy selection}$$

6:         Take action $a$ and observe $r$ and $s'$
7:         Store transition $(s, a, r, s')$ in $D$
8:         Sample random mini-batch of transitions $(\tilde{s}, \tilde{a}, \tilde{r}, \tilde{s}')$ from $D$
9:         Set

$$\tilde{y} \leftarrow \begin{cases} \tilde{r} & \text{if } \tilde{s}' \text{ is terminal}, \\ \tilde{r} + \gamma \max_{\bar{a}\in\mathbb{A}} Q(\tilde{s}', \bar{a}, \bar{\theta}) & \text{otherwise}, \end{cases}$$

10:         Perform stochastic gradient descend step with $\left(\tilde{y} - Q(\tilde{s}, \tilde{a}; \theta)\right)^2$   $\rhd$ DQN weight updates
11:         **if** $n = 0 \bmod(C)$ **then**
12:             Update target action-value functions parameter $\bar{\theta} = \theta$
13:         **end if**
14:         Update the state $s \leftarrow s'$
15:     **end while**
16: **end for**
17: **return** $Q(\cdot, \cdot; \theta)$

---

## 2.4.2 REINFORCE

While value-based methodologies focus on the learning of the action-value function to derive then an operational policy, the policy gradient algorithms adopt a completely different approach. In this case, the main idea is to directly approximate the policy. Specifically to the aims of this thesis, ANNs are applied for these estimations. As done for DQN, also here the weights of the ANN are identified as the parameters $\theta \in \mathbb{R}^d$ of the policy-approximator $\pi_\theta(s|a)$, $\forall s \in \mathbb{S}, \forall a \in \mathbb{A}$.

In order to update the policy, a performance metric $J(\theta)$ has to be selected. Given the fact that the value functions are capable of estimating the quality of a policy, the trivial choice is:

$$J(\theta) = V_{\pi_\theta}(s_{init}) = \mathbb{E}_{\pi_\theta}\left[G_0 | s_0 = s_{init}\right] \tag{2.20}$$

where $s_{init} \in \mathbb{S}$ is the expected initial state. From this information, the rule for updating the weights of the ANN can be derived following the gradient ascent method:

$$\theta_{i+1} = \theta_i + \alpha \nabla_\theta J(\theta), \tag{2.21}$$

where $\alpha$ is the step size or learning rate and $\nabla_\theta J(\theta)$ is the gradient of the state-value function evaluated in $s_{init}$. For calculating this gradient, the well-known *Policy Gradient Theorem* [231] can be used, which states that:

$$\nabla_\theta J(\theta) \propto \sum_{s \in \mathbb{S}} \mu(s) \sum_{a \in \mathbb{A}} Q_{\pi_\theta}(s, a) \nabla_\theta \pi_\theta(a|s) \tag{2.22}$$

$$= \mathbb{E}_{\pi_\theta}\left[\sum_{a \in \mathbb{A}} Q_{\pi_\theta}(s_t, a) \nabla_\theta \pi_\theta(a|s_t)\right], \tag{2.23}$$

where $\mu(s)$ is the proportion of times that $s$ was visited following the policy $\pi_\theta$. In other words, the Policy Gradient Theorem indicates the proportionality between the gradient and the time spent in each state multiplied by the value of each action scaled by the policy gradient. Moreover, this result can be rearranged through some algebraic steps. In particular, by dividing for the probability of taking the actions $\pi_\theta(a|s_t)$, then it is possible to replace the sum over the actions with the action $a_t$ sampled by $\pi_\theta$. Hence, it follows:

$$\nabla_\theta J(\theta) = \mathbb{E}_{\pi_\theta}\left[\sum_{a \in \mathbb{A}} \pi_\theta(a|s_t) Q_{\pi_\theta}(s_t, a) \frac{\nabla_\theta \pi_\theta(a|s_t)}{\pi_\theta(a|s_t)}\right] \tag{2.24}$$

$$= \mathbb{E}_{\pi_\theta}\left[Q_{\pi_\theta}(s_t, a_t) \frac{\nabla_\theta \pi_\theta(a_t|s_t)}{\pi_\theta(a_t|s_t)}\right] \tag{2.25}$$

$$= \mathbb{E}_{\pi_\theta}\left[G_t \nabla_\theta \ln \pi_\theta(a_t|s_t)\right], \tag{2.26}$$

where Equation (2.26) is derived by writing the derivative of a logarithm of the policy and using the definition of $Q_\pi(s_t, a_t) = \mathbb{E}_\pi[G_t | s_t, a_t]$.

The expression found in the expectation in Equation (2.26) is the necessary quantity that can be sampled at each time step for approximating the gradient. With this information, the stochastic gradient ascent update rule can be obtained:

$$\theta_{i+1} = \theta_i + \alpha G_t \frac{\nabla_\theta \pi_\theta(a_t|s_t)}{\pi_\theta(a_t|s_t)}, \tag{2.27}$$

that represents the main part of the *REINFORCE* algorithm [263]. This methodology takes all the future rewards up until the end of the episode into account. Therefore, it can be recognized as a Monte-Carlo approach that is well-defined only in episodic scenarios where the updates are made after completing the episodes. For the sake of clarity, the complete pseudocode of the REINFORCE algorithm is reported in Algorithm 2.4.

---

**Algorithm 2.4** REINFORCE algorithm [263]

---

**Require:** Differentiable policy $\pi_\theta(\cdot|\cdot)$, Learning rate $\alpha \in (0, 1]$, Number of episodes $N \in \mathbb{N}$, Discount factor $\gamma \in [0, 1]$.
1: Initialize policy parameter $\theta \in \mathbb{R}^{d'}$
2: **for** $n = 1, \ldots, N$ **do**
3:     Generate an episode $\{s_0, a_0, r_1, \ldots, s_{T-1}, a_{T-1}, r_T\}$ following policy $\pi_\theta$.
4:     **for** $t = 0, \ldots, T - 1$ **do**
5:         $G \leftarrow \sum_{k=t}^{T} \gamma^{k-t-1} r_k$
6:         $\theta \leftarrow \theta + \alpha \gamma^t G \nabla_\theta \ln \pi_\theta(a_t|s_t)$          $\triangleright$ Stochastic Gradient Ascent
7:     **end for**
8: **end for**
9: **return** $\pi_\theta$

---

To summarize, this chapter provided an introduction to RL's mathematical background and introduced the two major classes of algorithms applied throughout this work. This introduction contributes to delineating the foundations for the next chapter where insights on the branch of XRL are discussed.

# An Overview of Explainable Reinforcement Learning 3

In the following chapter, a structured overview of the most important advances in XRL in recent years is presented to identify various research streams. First, the dictionary necessary to avoid any misunderstanding derived from the use of unclear terminologies is reported. Thus, a common language needs to be established for the remainder of this thesis. Subsequently, using the key classification ideas noted in the literature review published in *Milani et al.* [161], review on advancements in this field is discussed. As previously described in Chap. 2, the citations to [161] are limited to increase the readability and clarity of the text. Finally, from the discussion of the limitations of state-of-the-art methodologies in XRL, the research questions addressed in this thesis are described and extensively reasoned.

## 3.1 Explainability and Interpretability: The Definitions in XAI

Defining a well-shaped dictionary for addressing the correct concepts is a relevant task, especially in the XRL field, where a multitude of terminologies are present. Inconsistencies in the definitions used in distinct studies from the literature can generate misunderstanding of the overall outcome and create a subjective perspective on the results [193]. In particular, significant problems are created when two works adopting contrasting terminologies are compared.

One of the flagship mix-ups in XRL (and XAI in general) concerns the clear description of the terms "Explainability" and "Interpretability". Indeed, a spectrum of papers are using those words as synonyms [164, 168, 193, 198, 238] while, for many others, there are major differences that distinguish them [13, 203, 268].

© The Author(s), under exclusive license to Springer Fachmedien Wiesbaden GmbH, part of Springer Nature 2026
R. Milani, *Advanced Automation for Comprehensible Causal Explanations of Reinforcement Learning Agents*,
https://doi.org/10.1007/978-3-658-50495-3_3

Another recurrent circumstance is when authors evade from explicitly giving the definition of what they identify as explainable or interpretable [137]. For the purpose of this thesis, the two nouns are treated differently. In general, the primary difference between these definitions consists of the output generated in the corresponding methodologies: while in "Explainability" the focus is on explaining the choices of the agent in human terms, in "Interpretability" the core objective is to unravel the inner mechanisms by interpreting its weights and features through specific measurements. Therefore, in the former case, the final output of the approaches are human-readable explanations to particular questions and, in the latter scenario, the attention is only on the evaluation of pure statistics and values. Thus, while the explainable methodologies are capable of answering target questions by providing users with simple and efficient reasoning that does not require advanced knowledge, the interpretable ones involve a more in-depth understanding of complex concepts, e.g., SHAP [131]. Hence, the intended audience of the corresponding outcomes is different: experts for the interpretable approaches and both experts and non-experts for the explainable methods.

Nevertheless, this ambiguity affects more than this single pair of terminologies in XRL. A large set of equivocal pairs can be listed, starting from "Transparency". This word has been used as a synonym of "Interpretability" [137] while in other situations, it is considered in completely different manners depending on the context [260]. Other remarkable nouns with an ambiguous definition are "Comprehensibility", "Simulatability" and "Decomposability" which are used always with different acceptions [83, 137].

However, due to the importance of these specific words related to the evaluation of the explanations in the successive chapters, it is fundamental to clearly define the following terminologies:

- *Transparency*: the ability to give relevant information about a model, e.g., feature relationships;
- *Trustworthy*: when a user can reasonably be confident in the choices of the algorithm;
- *Understandable*: if a user can comprehend why the explanation was generated in that way;
- *Satisfying*: the feeling of having all the information in a well organized manner;
- *Sufficient Details*: the feeling of having enough information;
- *Safe-reliable*: being able to constantly behave in the same way;
- *Predictable*: the possibility of an outcome to be forecast by a user;
- *Complete*: when all the fundamental components are presented in the explanation.

The choice of these interpretations for these nouns is based on the well-known paper by Hoffman et al. [103], which represents a milestone in the field of XAI. The comprehension of this vocabulary will be crucial for the correct evaluation of the results achieved in the human studies of Chap. 4. Nonetheless, terms relative to the computational performances of XRL have also to be addressed. In this case, when in the text is referred to the efficiency of a model, it will be related to the computational evaluations, i.e., the accumulated return obtained or the accuracy of the predictions.

In Chap. 5, the common adjectives utilized are "Important", "Fundamental", "Critical" and "High-Risk". They all describe principally the intrinsic qualities of a state. The first two are considered synonyms and define a piece of information that presents a high value for at least an importance metric. Although in Chap. 6 the mathematical definition of "Relevant state" is presented, it is necessary to clarify that it is different from the precedent terminologies. When reported as "Critical" and "High-Risk", it is intended to identify particular situations where making a wrong decision could lead to failing the final task.

Strictly connected to the importance of the states are the sub-goals or sub-tasks, which represent remarkable operations that have to be executed in order to have a positive outcome. However, it is fundamental to observe that not all of the sub-goals are bottlenecks, i.e., states where only a particular choice of action allows the agent to move towards the completion of the MDP. Some sub-tasks can indeed be optional. This kind of information is analyzed in depth throughout Chap. 5. However, it is necessary to present this concept now since it will be central for multiple studies discussed in the following review.

Despite this list of definitions not being exhaustive, it covers all the relevant aspects that will be encountered throughout the reading of this thesis. In the next section, the features studied for the classification of the XRL approaches discussed in the subsequent literature review are introduced.

## 3.2   The Novel Classification Scheme

The analysis of the methodologies in XRL follows some specific characteristics that are classically considered crucial. In the first place, the categorization is performed through the distinction based on two factors: *scope* and *time*. For the former quality, two possible types are used: *global* (G) and *local* (L). The global models are able to explain the overall logic behind the general behavior of an agent [1, 193]. On the other hand, local approaches target a particular decision from which the explanation will be generated [1, 126].

In the case of the time aspect, the classification is based on the moment when the explanation is generated: the models that are inherently interpretable or self-explanatory, e.g. decision trees, are called *intrinsic* (I); and the methods that need an auxiliary model to provide explanations for the original method are defined *post-hoc* (P) [126, 193]. More in detail, in the intrinsic approaches the explanations are directly generated through the algorithm without any additional tools, while in the post-hoc, there is always the requirement of a second methodology that can be coupled with the original RL agent. Therefore, the former ones can instantly generate interpretations of the selected choices and the latter ones have to wait for a diverse pipeline to create the descriptive outcome.

Another feature highly recognized for discriminating these methodological groups in XRL is the typology of the RL algorithm that can be coupled. Specifically, delineating whether it is *model-agnostic*, so, applied to any situation, and *model-specific* [3, 193]. In the forthcoming literature review, this analysis is not restricted only to selecting one of the aforementioned kinds but it describes the specifications of the appropriate RL algorithm adopted for each scenario.

These classical qualities used for characterizing each model are not sufficient in determining a clear distinction between multiple classes of algorithms. In fact, methodologies that share these factors can be applied in completely different contexts and they can use concepts that are unrelated to each other. Hence, in this proposed classification scheme, other necessary characteristics need to be introduced.

In particular, the grouping decided in this work is based on both the input and output features. Explicitly, by delineating the input data required, it is possible to cluster together methodologies that share similarities in the procedural pipeline. The same idea is applied for the categorization based on the output through the observation of the type of explanation generated, e.g., causal explanation or visualization of heatmaps. After this basic screening, the technical differences between the models employed can be easily categorized by analyzing their core ideas. Hence, the final step is strictly related to the description of the methodology. Through this schematic procedure, a total of eight classes, named from their methodological design, are determined in the literature review reported in the next section.

## 3.3   Literature Review of XRL

Even though the field of XAI is in rapid expansion, there is still limited knowledge of what is happening in the majority of the applications. This enormous inflation on the topic of explainability has also involved RL in recent years. This is attested by the number of publications with keyword "Explainable Reinforcement Learning", as shown in Fig. 3.1. Furthermore, this explosion of the number of scientific papers does not regard only the methodological improvements but also survey articles that try to find a coherent taxonomy in XRL [1, 37, 54, 65, 89, 90, 169, 196, 203, 209, 238]. The goal of this review is to provide a broad overview of the diverse subfields in XRL. Then, in the successive chapters, the specific settings required for understanding all the details of the proposed methodologies are discussed in depth.

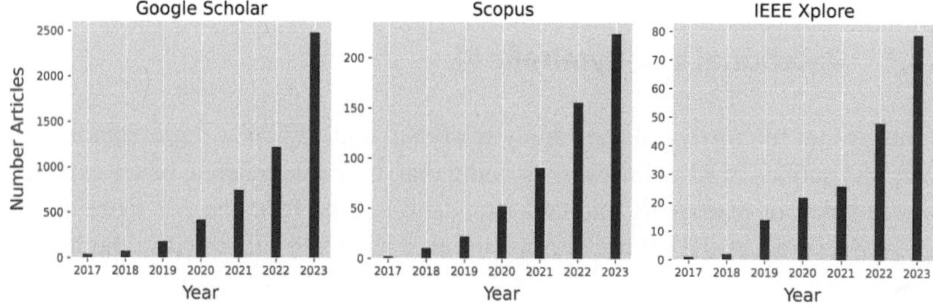

**Fig. 3.1** Bar plot of the number of articles found using the key word "Explainable Reinforcement Learning" in multiple search engines

For this reason, it is necessary to determine adequate boundaries in the selected literature to not overlook any area and be able to further analyze it more deeply. To this end, in this study, have been considered all the publications since 2017 that presented at least one of the following specific features:

- having at least 10 citations in any search engine;
- improving previously described methodologies;
- introducing an innovative idea, i.e., diverse from the established approaches.

The choice of introducing the last two conditions is justified by the presence of relevant studies that have been only recently published and that would not be, at the moment, well-known by the XRL community. In this way, the new promising publications can be included in the corpus of the analysis.

The comprehensive collection of papers was compiled by querying three major academic search engines –Google Scholar, Scopus, and IEEE Xplore– using the following keyword combinations: "Explainable Reinforcement Learning", "Interpretable Reinforcement Learning", "XRL", "Explainability in Reinforcement Learning", "Explainable Reinforcement Learning Visualization", and "Causal

Explainable Reinforcement Learning". From this initial set of resources, backward research was performed in order to avoid leaving behind any critical methodology that has not been found using the preceding keywords. From this collection, the papers that did not satisfy the aforementioned properties have been discarded. The remaining ones are grouped into eight classes generated from the classification scheme discussed in Sect. 3.2.

In the next subsections, the details of each group are discussed, highlighting especially the benefits and the pitfalls of the relative methodologies described.

### 3.3.1   Relational and Symbolic RL

In this group, the methodologies apply relational and symbolic representation for states and actions in RL. However, by doing that, they can generate better explanations in terms of readability and understandability. The focus here is more on the logical encoding of the salient information that has to be presented to the human users.

One of the first and principal papers dealing with this kind of idea is [269]. The authors introduced the concept of Relational Reinforcement Learning in the DRL context to increase the efficiency, generalization capability and interpretability of deep learning methods. The general approach in Relational Reinforcement Learning consists of translating states, actions and policies using first-order (or relational) language. Due to this particular construct, it is possible to directly include background information such as well-known rules related to the problem that can be encoded by a logical fact. Specifically for the deep learning scenario, the authors created a complex architecture consisting of convolutional layers and a relational module composed of multi-head dot-product attention (MHDPA) [246], or self-attention, blocks. In terms of results achieved, this approach showed good generalization capability. Nonetheless, problems with overfitting in the training phase arose when the dimensions of the architectures increased.

Following the previous idea, a symbolic representation of DRL was applied in [145] to improve its interpretability. This led to the creation of a new framework defined as Symbolic Deep Reinforcement Learning and comprising three components: a planner, a controller, and a meta controller, each of them fulfilling a particular task. For the planner case, it has to plan, through a prior symbolic representation [132], what actions the agent has to perform in order to achieve the final goals. The controller, instead, has the role of finding the solution of the MDP, thus, it is a classical DRL method; DQN in the analyzed paper. However, the reward used in the training of the controller consists of two components: the *intrinsic* reward, which is the usual one determined by the environment, and the extrinsic reward, learned through the meta controller. The latter measures the performances of the controller in suggesting different intrinsic goals. With these convoluted interactions between the three parts, the learning of sub-policies for each sub-goal completion is facilitated. Moreover, due to the task of the meta controller, it is possible to identify the sub-goals with higher accuracy. The computational time required for the training of the overall approach is quite expensive. In [145], although the experimentation was performed in simple environments such as Taxi and ATARI games, more than a million episodes were necessary for identifying all the sub-policies and sub-tasks. Furthermore, not all these sub-goals were successfully learned by the controller.

Improvements to the previous procedure have been developed in [146], where a Neural-Symbolic Reinforcement Learning approach is proposed. Also here, three main components are utilized to enhance the transparency and interpretability of DRL methods. The first one is the same as in [269]: an attention module MHDPA is employed in the process of generating attention weights on logical rules with different lengths from the symbolic states. The symbolic states represented as matrices and the resulting weights from the initial step are given as input to a reasoning module which produces reasoning on the actual information. Finally, the results obtained by the precedent modules are elaborated by a Multi-Layer Perceptron (MLP), called the policy module, to individuate the action predicate (symbolic representations of the actions through first-order logic). Hence, at each time step, the selected action is based on the state-action values of the action predicate matrices. The authors state that, with this type of methodology, it is possible to increase the interpretability of the chosen action and generate explanations through visualization of the relational paths to different sub-goals. In particular, an interesting future direction can explore the application of trees and junction rules to improve the explanations through more expressive rules.

Another similar approach called NUDGE is discussed in [60]. The key characteristic of this method lies in its ability to convert the raw states into logic representations, obtaining entities and their relations useful for deducing the agent's actions. The explanations provided indicate which input is important using the gradient-based attribution algorithm over logical representations [230]. Computational tests demonstrate that NUDGE can compete with neural-based networks such as PPO, and adapt to the modifications in the environment. Even though the efficiency tests achieved remarkable results, no user evaluation, that could determine whether the logical reasoning was sufficiently expressive, was conducted.

## 3.3.2  Reward Decomposition

As can be noted, the approaches investigated above presented numerous advanced techniques combined to achieve outstanding results. Some core ideas are also shared between methodologies in different classes. An example from the previous section is [145], where a reward decomposition was implemented. Therefore, the group of algorithms described here are the ones dealing with these specific tactics.

The Reward Decomposition, a technique introduced in [206], is adopted to branch the classical reward to extract more information from the learned policy. The main issue deriving from this process is the choice of reward shaping. Indeed, while in various environments there is a natural reward decomposition, this can not be generalized, leading to moving the importance of the findings to the specific consideration assumed in the reward division.

Similarly to the previously disclosed relational and symbolic representations, the concept of reward difference explanations, based on the definitions of semantically relevant rewards, is introduced in [69]. Associated with each of the decomposed rewards are the corresponding Q-functions, which can be summed up to generate the full-knowledge state-action value function. Thus, it is possible to train a tabular method considering only the specific kind of Q-function. In [69] the RL algorithm considered was SARSA. Despite its simplicity, it showed a high level of efficiency and understandability. However, in their experiments, the explanations obtained by the authors heavily rely only on one reward, consequently unbalancing the description towards that feature and discriminating the overall evaluation of the other factors.

A later improvement to this concept is proposed by [118]. In their work, the authors showed an off-policy algorithm capable of learning the decomposed state-action value functions in a way that the sum converges to the optimal value. In fact, in the previous methodology, learning each decomposed Q-function does not guarantee that the sum will converge to the correct full-knowledge value function. Thus, a modification in the updates of the decomposed state-action value function is necessary. Therefore, in [118], it is suggested to first evaluate the action that maximized the summed value function and then modify each corresponding decomposition following the usual formula of Q-learning.

The aforementioned approaches generate explanations based only on the single step where the agent resides at the moment. This aspect can be misleading since it does not take into consideration any future events or possible sub-goal planning. In [199] this flaw was identified and, consequently, corrected by combining the classical reward decomposition with a hierarchical technique. The derived framework is composed of two DQNs, one for a higher-level task and the other for a lower-level task. Following [129], the higher-level DQN aims to identify the correct sequence of sub-goals necessary to solve the MDP, while the lower-level DQN learns the policies for attaining the actual sub-goal. The latter DQN algorithm is trained using the deep learning version of the previous reward decomposition updates. In this way, the single-step explanation can be coupled with the explanation relative to the planning for completing the respective sub-goal.

In all the case studies presented, no human evaluation has been conducted to assess the explanations generated using reward decomposition. Taking into account that the final ambition is to increase the transparency of complex RL algorithms, estimating the effects that those descriptions have on users and understanding how humans perceive these justifications are major aspects to bear in mind.

Since the general notion of the hierarchical method has already been stated while discussing the paper [199], in the next section the class treated is the hierarchical and goal-based one.

### 3.3.3   Hierarchical and Goal-Based Methods

Since the general concept of the hierarchical method has already been described while discussing the paper [199], this section focuses on the hierarchical and goal-oriented approaches.

The hierarchical and goal-based procedures are determined by the interaction of multiple-level policies. The simplest scenario consists of only two layer policies: The first layer policy is applied for the recognition of the sub-goals, while the second layer provides the optimal solution to achieve the previously defined task [232].

The chronologically first approach in XRL to include this kind of process is presented in [219]. In this paper, a hierarchical structure has been adopted to generate more interpretable policies from actor-critic agents. The pipeline introduced there considers a multitasking RL where each sub-goal is identified using two-word templates: one for the skill that has to be learned and the other for the item involved in the assignment. To train an agent in the full environment, the dimensions of the set of sub-tasks are progressively increased. With this choice, the RL algorithm initially focuses on learning only to achieve the fundamental sub-goals. Then, it exploits the learned skills to tackle more advanced situations. The hierarchical procedure utilizes four distinct policies:

- a base policy for the already learned tasks;
- an instruction policy helpful in allowing the actual policy to communicate with the base policy;

- an augmented policy that allows the actual policy to perform the actions;
- a switch policy deciding which policy, between the base policy and the actual one, consider for the next time step action selection.

The direct choice for a base policy is the previous time-step trained policy. The main disadvantage is the necessity of a human expert who can decipher the environment and recognize the correct tasks to be learned. In the experiments performed in [219], the goals have been selected through human instruction. On the other hand, the proposed hierarchical plan achieved positive results both in terms of efficiency and transparency.

A different goal-based method is presented by [29], which introduced the *Dot-to-Dot* algorithm. The methodology implemented results from the combination of various ideas before the classical hierarchical plan exploitation. In detail, in order to solve the MDP, a Deep Deterministic Policy Gradients algorithm [136] is trained using the Hindsight Experience Replay [10] helpful to rescale every trajectory importance in the training phase regardless of the achievement of the goals. The improvements in the two-level hierarchical structure are in the formalization of the sub-goals. Specifically, limiting the analysis for sparse reward environments, the sub-tasks that have to be learned are small perturbations of the previously completed goals in the same episode. By promoting the attainment of nearby tasks, it is possible to evaluate the importance of each action-state couple. Nevertheless, it neither provides any information regarding the choice performed nor improves its trustworthiness.

### 3.3.4  Consequence Based Methods

Up to this point in the literature review, only a few methods considered the future consequences of the chosen actions in the creation of satisfactory explanations. Nonetheless, this can be a relevant factor for convincing the users of that particular selection. Indeed, multiple ideas have been developed in this class, as noted in this section.

The consequence based explanations rely on the comparison of the future modifications due to the actual choices of actions. The general idea of this class is based on how usually people explain particular decisions: through contrastive questions that compare the collected results [164].

In [253], a first improvement to the classical approach proposed by [95] has been developed. Initially, the states are transformed into interpretable variables through an appropriate function. Thus, the rewards accumulated are justified through the outcomes transformed by the previous function. The novelty brought by [253] consists of using a policy Q-value function generated from the combination of two value functions: the classical trained Q-value function and a preference function used for discriminating the user preference in selecting each action. The final explanation is created from the comparison between the trajectories stored by the RL algorithm applied and the foil policy produced by the foil state-action value function. The main advantage of this procedure is being completely model-agnostic. However, a major pitfall is linked to the choice of a user preference function. In fact, various people can have completely different perceptions of what is better on the same occasions [210]. Therefore, the particular preference function adopted can generate biased explanations that are not trustworthy in general.

Instead of evaluating the preferences, it has been proposed by [56] to calculate the risk associated with each state. This measure is computed by considering the next $n$ states visited by a trained agent and estimating the corresponding feature importance. From this study, the authors observed that not all the characteristics present the same usefulness in determining a satisfying explanation of RL models. Therefore, the information relevant to the calculation of the risk function has to be part of a limited subset of features. Nevertheless, this particular selection is strictly dependent on the environment involved. Despite being an efficient post-hoc model-agnostic approach, it depends highly on the definition of the risk, which, in turn, can be safely assessed only by experts.

In the previous cases, the value functions, and the Q-value function in particular, played a fundamental role in discriminating the differences between the outcomes obtained by a specific action. Nonetheless, the availability of this information is not always verified. Thus, it is interesting to describe the complex framework from [91] which does not depend on the state-action value function. In this paper, the authors proposed to replace the value functions with the final reward of an episode. This prediction can be achieved by combining Recurrent Neural Networks (RNNs) with a Multi-Layer Perceptron (MLP) to embed the trajectories' information. Additionally, a Gaussian Process is employed to generate a latent representation of the episode, which is then used for linear regression to predict the final accumulated reward. The explanations produced consist of the sequence of the most important states in the episode leading to the forecasted return. The evaluation process adopted can not be recognised as rigorous since they considered a fidelity interpretation: comparing the most critical time steps derived in different interpretation methods.

### 3.3.5   Causal Models

The modifications of future events can be recognized as causally influenced by the choices of the actions. Thus, these consequence based models that rely on the causality can also be classified in a more specific group: the causal models.

The causality is one of the principal factors considered by human users for explaining their decisions [184]. This motivates the reasons why multiple methodologies try to replicate the same process in XRL.

A well-known example is proposed in [149]. Here, the authors utilizes an action influence model as a post-hoc approach for explaining model-free agents. This particular causal graph has the property of summarizing the state features as nodes and actions as edges. Through this structure, the explanations are generated looking at the chain of influence of the selected action. Moreover, classical ML models are employed for the prediction of the chosen action, making this approach easy to implement. The efficiency and reliability of this methodology are tested on simple environments reporting satisfactory results. Also the understandability is evaluated in a human study obtaining good results in terms of completeness and sufficiency.

The same authors improved this idea in [148] by introducing the concept of distal actions, i.e., actions that have to be enabled by a precise sequence of previous steps. In order to forecast this information, they adopted RNNs with input the trace of the episodes. Furthermore, they included in the comparison of the ML prediction models also the decision policy trees and proved their top efficiency. Although the prediction accuracy achieved is high, a main limitation concerns the construction of the action influence model. At the moment, there are no known algorithms that can discover the correct action influence model, while there are methods that can learn causal graphs automatically [180].

These methodologies are the starting point of the methodologies proposed in this thesis. Therefore, they will be discussed in depth in Chap. 4.

Coming back to the classical causal graphs, [99] discovered the causal representation in diverse RL environments by applying the mediation analysis [6, 187]. This methodology allows to compare two policies, one trained to successfully solve the MDP, and the other to fulfil the truthfulness of a particular variable. In this way, it is possible to identify the causal importance of that variable in reaching the final goal through the computation of the Natural Indirect Effect [81]. The proposed causal graph satisfies the property of parsimony, restricting the insights on fewer important features. However, this condition is not always verified. Generally, MDPs can be very complex, presenting relationships between variables and goals difficult to summarize in a few characteristics. The sparsity condition on the features used in the causal models has also been considered in the work of [250]. Despite improving the understanding of the environment, this approach did not provide insights into the behavior of the agents. In addition, neither of the two previous studies take into account the problem of causal confusion: the phenomenon associated with the generalization of relations between features that do not hold in every situation.

For this specific problem, [79] introduced an algorithm to correct this aspect before deploying the explanation of the agent choices. The methodology is based on the creation of alternative environments from critical states of a particular policy in order to evaluate the performances compared to a different policy taking into account only a smaller set of features. This expedient enables the identification of the previously crafted environments where causal confusion is detected. From the analysis of these particular situations, it is possible to discriminate the correct subset of features to avoid this problem and let the starting policy act differently. The human user can have a better overview of how the features are connected to the sub-goals with the adoption of a policy dealing only with a subset of states' characteristics. Nevertheless, the differentiation of the critical states has to be supervised by a human expert. Thus, this kind of method can be less biased on the human perception in the case of the automation of the aforementioned step.

The practical capabilities of causal models is that they can be coupled together with multiple methodologies, without losing their main properties and qualities. A recent paper [255] combined the usage of particular structural causal models with perturbation-based importance measures for the creation of situational aware explanations. Precisely, they proposed to fix the causal structure found for a time step and then repeat it multiple times to predict future outcomes. The resulting model has been called the cascading structural causal model due to its repetitive shape. The application of the importance measure is fundamental for recovering the information to answer counterfactual questions. Computational results show the efficiency of this technique. Nonetheless, no user study has been performed, hence, the qualities of the generated explanations were not graded. Furthermore, the structural causal models can present great difficulties in detecting complex relations in high-dimensional scenarios.

Similarly, [143] merged the classical structural causal models satisfying sparsity, sufficiency, and orthogonality conditions with the reward decomposition techniques. Through the definition of specific importance metrics, the aforementioned properties were optimized during the search for the optimal structure. Moreover, they also included an analysis of the critical states through the Q-value-based measure proposed by [239]. Although this complex framework can provide efficient solutions compared to other intrinsic methods, it also assumes the existence of a differentiation of the rewards in the environment. Therefore, this procedure can be applied only in environments presenting multiple tasks.

It is relevant to notice that only a few studies consider increasing the explainability of RL agents through causal models [267]. The majority of the approaches that combine causal models and RL try to improve the general capabilities of the agent without focusing on the possibility of also enhancing the comprehension of the users that will deploy these algorithms. Additionally, causal reasoning is a fundamental of human decision-making, its utility in aiding decision-making is nuanced and may depend on the individual's prior experience and the complexity of the decision being made [271]. Hence, further research is needed to develop methods that can effectively utilize causal models to enhance decision-making across different contexts and levels of experience [124, 271].

### 3.3.6  Human Robot Collaboration

As noticed in [79], human help can give interesting insights for determining useful information. However, a fundamental necessity is represented by the elimination of any user errors. Therefore, it is fundamental to balance the human intervention and the computational power to find an optimal solution. In this section, procedures that try to accomplish this goal are outlined.

In modern times, the collaborations between humans and machines have increased bringing advanced solutions to complicated problems but, at the same time, increasing drastically the number of potential errors caused by misunderstandings [244]. In the methodologies reported hereinafter, the theoretical framework considered is a subclass of MDPs named Partially Observable Markov Decision Processes (POMDPs) [231]. In this particular scenario, the agent is allowed to gather only a subset of the full information from the environment, leading to further complicating the tasks to complete.

Chronologically the first approach that dealt with filling the gap between humans and robots is [194]. In this study, an algorithm is developed that generates an explanation through a combination of the principles of Situation Awareness - based Agent Transparency (SAT) [42] and Natural Language Processing [175]. From these concepts, four different classes of explanations are derived and tested in a human evaluation study. The first one is the simplest: the none explanation, where no information other than the selected action is provided. Then, the level of information provided to the testers increased with the number of the class. The tested users evaluated the last two classes' explanations as being more reliable in helping them to recognize possible wrong decisions. Moreover, from this analysis an interesting observation can be expressed: it is reported that the design of these explanations is relevant to the human mind. The descriptions from the second class were perceived as good as not having any other information. Therefore, a correct identification of the data necessary for increasing the trust of the users is mandatory.

Another problem that has been addressed is the correction of erroneous actions characterized by the evaluation of incomplete information learned from the reward function. To fix this issue, [234] introduced a method for the generation of diverse reward functions used for the comparison of the results achieved by the RL agent. In this process, the human knowledge is fundamental for the identification of the initial reward functions applied as tests. Moreover, the corrections are drawn from a new policy, crafted through the previous analysis, that chooses when to intervene. Despite proving positive overall experiences in increasing the trust in those corrected policies, the set of possible reward functions that can fit the trajectories in the memory is exponentially exploding. In fact, to avoid this obstacle, only trivial reward functions have been included in the study.

In [50], the authors leverage the level of expertise of the users in specific environments to produce the best explanations fitting their knowledge. They focused their attention on two kinds of local explanations: one technical and the other shaped to be more human-like. In detail, the former approach is based on the Q-value associated with the current state and the selected action. On the other hand, the latter approach involves reasoning through the analysis of the probability of successfully completing the task. The human users involved in verifying the qualities of this methodology confirmed the superiority of the human-like description. Furthermore, only a small group of experts evaluated the technical expressions as the best. Nonetheless, this experimentation was performed in a modest scenario, with a simple environment and just two kinds of explanations. To completely check the maturity of this approach, an extended analysis of a complex problem with multiple sub-goals has still to be carried out.


### 3.3.7   Visual Based Methods

Whilst in the previous section the focus was on bridging the gap between the human perception and the computational learning process, in the class discussed here, the attention is moved to methodologies that exploit visualization techniques to profitably explain the RL algorithms' choices.

The most prominent techniques applied in XRL delve into visual representations to mediate the human understanding of the agent's actions. This popularity is linked to the ability of visualizations to increase trust and interpretability for black-boxes ML algorithms [198].

In this class, a particular subgroup can be recognized as the most prolific one: the saliency maps methodologies. Saliency maps are peculiar images presenting highlighted regions in order to facilitate the comprehension of where to focus. The first and main contribution to this concept is proposed by [87]. In this paper, the authors focused on the explanations of actor-critic agents through the evaluation of saliency maps of both the policy and value functions. For the creation of these maps, they evaluated the value function in a sequence of images and their perturbed versions. Afterwards, the squared magnitude of the difference between the previous results led to the importance metric of the map. The same procedure was carried out using the policy for the evaluations. From this initial methodology, multiple ideas sparked, such as slightly modifying the saliency measures by adopting a soft or binary mask for the perturbations [266], or using, instead of the policy and value function, the Q-values [112]. In [87, 112], the saliency maps were judged in a human study, gathering both experts and non-experts. The results obtained in this analysis confirmed the positiveness of this methodology, specifically in terms of understandability but also for the accuracy achieved.

From these preliminary studies, [92] proposed a more complex formulation for the saliency metric. The authors took into account not only the softmax of the action-state value function but also the Kullback-Leibler divergence [130]. With this modification, two properties are fulfilled:

- specificity: focus on the effect of the perturbation for the given action that has to be explained [92];
- relevance: the alteration in the expected reward should be lower for actions different from the considered ones [92].

A common fashion of creating novel approaches in XRL is to transpose classical XAI algorithms to the RL world. An example is [107], where the saliency maps are based on the layer-wise relevance propagation method [15]. The improvement contributed in the paper concerned the adoption of an $argmax$ implementation of the aforementioned approach to produce sparse maps. Another case of methodologies translated from XAI in general to the specifics of XRL is [117], where the Grad-CAM [211] has been employed. In this context, the procedure for the production of the saliency maps is the following: first, the weighted gradient for each node in the Convolutional Neural Network (CNN) is computed, then the global average pooling is applied. Finally, these results are combined with the feature maps and given as input to an activation function to output the measures. Although being a theoretically well-designed process, the explanations that are created are not easy to understand in the case of non-experts in ML due to the presence of multiple accentuated causes. A further example in the generation of saliency maps with XAI approaches consists in the integration of the concept of SHAP values [144] in XRL scenarios, as implemented in [256]. The main difference to the classical algorithm is in the evaluation of the actual state and its perturbation. Here, the contributions of the background data are computed instead of calculating the ones of the input's features. This method has been also tested in practical problems like in the traffic signal controls [200], obtaining improvements in the explainability and in the elimination of some biases related to the training phase.

As confirmed in the discussion of the preceding papers, the saliency maps help the users increase their trust in the RL policies through simple but effective visualizations. Nevertheless, some pitfalls are not addressed, e.g., the multiple influences highlighted in the input or the complex pipelines that could degenerate, in general, the comprehension. Therefore, in the next paragraphs are reported other visualization methods that try to counterbalance the explicated issues.

In [254], a comprehensive visual analysis for DQN agents is constructed. The visualizations showed various levels of detail, from the overall training trend to the epoch or episode overview. The interesting contribution of the authors lies in the possibility of analyzing segments in the agent's trajectories. This particular task is achieved through the combination of a dimensionality reduction Principal Component Analysis (PCA) and the non-linear method t-SNE [147]. This complete approach resembles a visual multi-criteria Decision Support System for specific RL algorithms. Given this particular structure, it remains difficult to generalize it in complex environments, especially in MDPs where the action space has large dimensions, e.g., Montezuma's Revenge Atari game. Furthermore, displaying multiple features can complicate the clarity of the explanation. More recently, also [236] suggested to leverage on the visualization of abstracted trajectories to help non-experts in RL understand the behavior of the agents. To this end, they created an interface that shows two main pieces of information: a map view identifying the trajectories as a directed graph and a slider for the representation of the trajectories over time. The technicalities used for the trajectory abstraction were based on the combination of particular Variational Autoencoder [93] and a clustering algorithm ST-DBSCAN [39]. The quality of the visualizations is proved in a small (only 9 users surveyed) study, finding that the slider data is more useful for the non-experts than the map.

The pitfalls of the saliency maps are pointed out also in [53, 165], where the authors tried to replace the view of the salient pixels in the input images with the important moments in the training. Specifically, during the training of the agent, the snapshots helpful for understanding the choice of the action are stored. This task has been completed by applying a Sparse Bayesian Learning algorithm (the Relevance Vector Machines[237]) that can pick the correct images and use them as an explanation. The sparse set of salient moments reduces the required memory and simplifies the interpretation but relies only on a small amount of information. Hence, this model and its sparsity can present relevant difficulties in complex environment applications, e.g., continuous state-action spaces.

The same idea has been applied in [212], where the authors extracted the most interesting information but in the form of video clips. In this way, the key moments of the trajectory can be summarized using a few interesting elements. As already stated in the previous papers, also in this case the real-world exploitation of this methodology can show uncertainties in relying only on specific features considered. Moreover, the theoretical background required finds a reliable basis in the work produced by [7], which is analyzed thoroughly in the next section.

### 3.3.8  Policy Simplification

In this final section of the literature review, the class of policy simplification methods is analyzed. Methodologies belonging to this group share the core idea of summarizing the most important information of the trajectories to let the user focus on a few, significant data.

An interesting example is given by [7], where the authors considered the usage of a particular Q-value metric for the recognition of relevant instants in the trajectories. For a fixed state, the value calculated consists of the difference between the maximum and minimum value of the state-action value function. In this way, the computed value provides a useful measurement for how much a state is important [239]. The results shown in an human study reported good results in improving the understanding of agent selection through preference summaries. From this analysis, a large number of metrics were developed to identify the critical states. In fact, including this kind of information in the explanations can increase the trust in RL agents and allow humans to feel safer [106, 239]. These particular metrics are investigated in depth in Chap. 5, illustrating the multiple facets of the aforementioned measurements. The HIGHLIGHTS method proposed in [7] was compared in a user study to the ASQ-IT pipeline consisting of applying Linear Temporal Logic on Finite Traces to generate explanations for non-experts [9]. The results reported better performances of the latter method in helping the users to correct possible mistakes and an overall higher explanation satisfaction.

A different way for approximating the Q-function is the trees methodologies. This subgroup is also widespread and very popular due to its simplicity in both interpretations for the human users and the implementation. Moreover, strong theoretical results are reported in the literature, providing basic foundation for multiple advancements in terms of explainability [41]. In addition, the authors in [139] adopted the ability of trees to efficiently approximate continuous functions by introducing the Linear Model U-trees extending the Continuous U-Trees, presented already in [243]. The particularity of their structure resides in the application of a linear model in each leaf node. With this construction, the accuracy and interpretability of the tree are increased. However, the obtained explanations are understandable only in scenarios where the depth of the analyzed tree is small. Thus, this methodology can be employed only in low-dimension problems.

Along the same lines, [220] modified the classical decision trees introduced in [229] to make them differentiable. The modification introduced consisted of replacing the boolean decisions with the use of sigmoid functions. In this study, the evaluation of the descriptions generated has been carried out in a small human study with 15 participants. Despite the limited dimension of the analysis, significant differences in the usability of these models compared to full MLPs were shown.

A similar approach that tries to reduce the information from bigger models to small trees is provided by [78] where the use of Soft Decision Trees was proposed. Here the changes brought consist of the combination of the classical tree model with binary splits and neural networks. In [48], the aforementioned algorithms have been applied for summarizing the deep learning policies, rather than the state-action value function as done by the previously mentioned methods. Using this idea, they achieved a prediction accuracy comparable to DNNs. For the explanation, the authors considered the rendering of heatmaps created from the selected features in the tree. A fundamental aspect that has to be fixed at the beginning of the experiment is the maximum depth of the tree. Despite this condition being highly determinant for the final positive results of the overall process, it is difficult to generalize a rule for its choice.

Instead of mixing decision trees and supervised learning techniques, [245] proposed the Mixture of Expert Trees based on the summarizing concept of Mixture of Experts [113] and on the CLTree clustering method [138]. Furthermore, the pipeline included also a critical state identification through the already introduced Q-function metrics [106, 239] helpful for the production of differently sized explanations [151]. Equivalently, [43] trained an ensemble decision tree using principally the teacher-student Q-metrics from [239] to produce post-hoc explanations for both local and global occasions.

Distilling policies can be aided by other optimization methodologies. A relevant example is [97], which applied the Particle Swarm Optimization [121] to create interpretable policies. In particular, a set of initial minimal rules is defined at the beginning of this process. Subsequently, this set is increased by the introduction of new rules according to the performances achieved. In the end, the best policy, i.e., the one that attains the highest reward with the minimum number of rules, can be detected and used to discover insights into the problem. This procedure is tested in systems dynamics problems reaching high-quality results in small environments but mediocre performances in real-world simulations, where a higher number of rules is required for increasing the comprehension.

Following the idea of using a well-known optimization methodology for the identification of beneficial policies, the same authors implemented a genetic programming version of the previous algorithm [98]. In this case, the final output is composed of an explanation in boolean terms and compact algebraic equations. The test results attained in real-world problems show better capabilities of this approach when compared to full NNs. Similarly, also [248] applied genetic programming to determine symbolic understandable rules for specific multi-objective problems with delayed rewards. The identification of lighter solutions that can be comprehended was allowed by the use of an upper confidence bound for determining which policy to simulate next. Genetic programming can also be coupled with the decision tree structure to improve the efficiency of the final explanation tree obtained, as done in [51]. Later in [72], the authors improved the previous pipeline by integrating an innovative component utilizing the Covariance matrix adaptation MAP-Elites [73]. The addition of this complex methodology improves the quality-diversity of the trees. In this way, the convergence speed is significantly improved due to the limited "illuminated" (i.e., balance between exploration and exploitation) search space. Recently, the number of algorithms based on advanced decision tree formulation is increasing quickly, and also their applications in multi-agent RL context started to spread [163].

Creating boolean rules for summarizing the policies, as mentioned in [98], is a direct way of explaining the corresponding decisions. Therefore, [152] proposed a post-hoc approach applying the boolean Decision Rules model presented in [55] for creating explanations of Rainbow DQN agents. The accuracy in the agreement between the agent and the rules was on average about 0.83, leading to the successful completion of the lava gridworld problem. Instead of using the boolean logic explanations, [105] utilized a more complex model including interpretable fuzzy rules for RL. Specifically, through the fuzzy system AnYa [11], it was possible to approximate the state-action value function, thereby identifying the favorable states.

Another approach used for simplifying policies is provided by [247], where the Neurally Direct Program Search for finding optimal policies related to a predefined syntactic one is described. The proposed algorithms guide a direct policy search to find the easiest policy to understand. By doing so, the explanations that are produced are human-readable and understandable. Nevertheless, the efficiency of the derived policy is undermined. The same idea was applied in [19], where the policies were found following an Imitation Learning process (a behavioral cloning process) [16] guided through Q-value metrics for important states detection. Although this approach improved the quality of the policy generated, it presents high costs in terms of time and computational power required to find the solution. Additionally, this pipeline was tested only in simple environments, thus, it is beneficial to evaluate it in complex situations to establish its real-world applicability.

## 3.4  Discussion and Detailed Formulation of Research Questions

From the reported literature review, it is clear how a vast majority of studies involved the combination of multiple methodologies. Many times, the algorithmic techniques employed also belong to completely different fields as in [143, 236, 245]. For this reason, it is impossible to theorize a taxonomy based on the particular group of concepts used. In fact, the specific characteristic of a taxonomy lies in the absence of intersections between different classes, which is impossible in the XRL case, as shown in Fig. 3.2. Due to this fact, the aforementioned classification scheme, which groups the methodologies that share the core idea and the shape of the output explanation, was proposed.

From Table 3.1 unexpected insights can be drawn. In the table, novel ideas and their associated properties are listed. Specifically, it includes a detailed description of the kind of methodology proposed, not only in terms of time and scope but also comprising qualities like the RL agent or data considered in the analysis, the type of explanation generated, whether or not it has been tested in a user study and the level of automation of the overall process. Focusing on the latter two aspects, it is surprising how few of the surveyed papers (30%) are carrying out human evaluations for their explanations. If then the aforementioned quality is considered together with the property of using an automated pipeline, the number is reduced to 20%. Just one out of five paper is, simultaneously, giving importance to creating a procedure that does not imply human intervention and being tested in a user evaluation. However, the combination of these features is an essential requirement for the practical exploitation of the suggested methods. Otherwise, in absence of the automation, the algorithms applied can be biased by the specific human perception used for its training, and, when not carefully tested by users, it can produce useless explanations that do not improve the general understandability or the transparency of the choices.

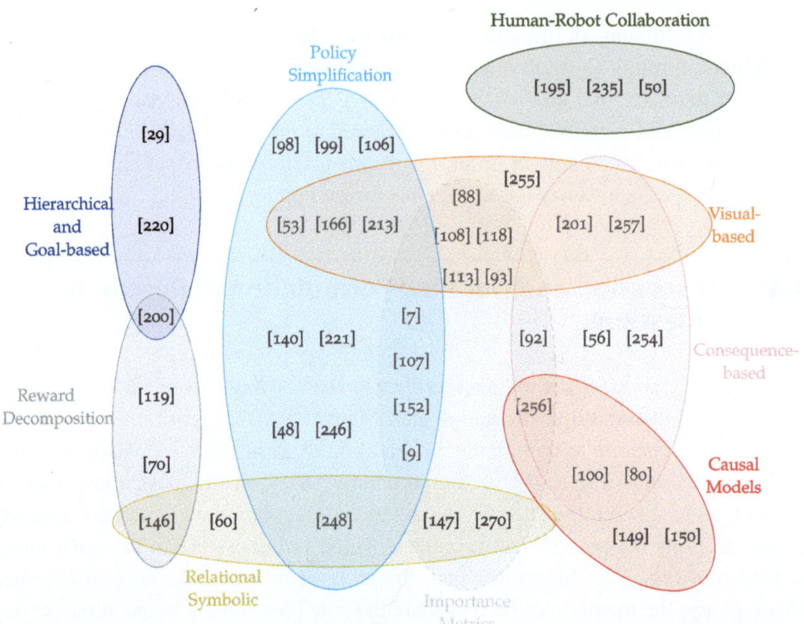


**Fig. 3.2** Venn diagrams representing the classification groups shown in the literature review. It is included in dashed line the group of Importance Metrics, due to its relevance in Chap. 5 (Modified from: Advances in explainable reinforcement learning: An intelligent transportation systems perspective, Edition by Rudy Milani, Maximilian Moll, and Stefan Pickl, Copyright ©2023 by Imprint. Reproduced by permission of Taylor & Francis Group)

The lack of methodologies that are automatic and user-validated is particularly significant for specific classes such as the causal models, where no proposed approaches satisfied both requests. Despite causal reasoning having a central role in human decision-making [271], a comprehensive procedure for generating causal explanations for RL agents is missing. From this shortage, the first research question is formulated:

**Research Question 1 (RQ1):** How to automatically generate causal explanations that can be human-validated?

**Table 3.1** Selection of the principal papers for the reviewed literature in the field of XRL following the proposed classification rules. The symbol ~ indicates that it is not clear if that property holds (Modified from: Advances in explainable reinforcement learning: An intelligent transportation systems perspective, Edition by Rudy Milani, Maximilian Moll, and Stefan Pickl, Copyright ©2023 by Imprint. Reproduced by permission of Taylor & Francis Group)

| Reference | Year | Time | Scope | Agent | Explanation | Human | Automatic |
|---|---|---|---|---|---|---|---|
| [269] Zambaldi et al. | 2018 | I | G | AC | Graph/Importance | × | ✓ |
| [145] Lyuet al. | 2019 | I | G/L | DQN | Symbolic Policies | × | ✓ |
| [60] Delfosse al. | 2024 | I | L | PPO | Symbolic Policies | × | ✓ |
| [146] Ma et al. | 2021 | I | L | DQN | Importance | × | ✓ |
| [69] Erwing et al. | 2018 | I | L | SARSA | Values/Bar chart | × | ~ |
| [118] Juozapaitis et al. | 2019 | I | L | Q-learn./DQN | Values/Bar chart | × | ~ |
| [199] Rietz et al. | 2022 | I | G/L | DQN | Values/Bar chart | × | ~ |
| [219] Shu et al. | 2017 | I | G/L | A2C | Hierarchical Graph | × | × |
| [29] Beyret et al. | 2019 | I | G | PG | Heatmaps | × | ✓ |
| [253] van der Waa et al. | 2018 | P | G/L | Value-based | Consequences | ✓ | ✓ |
| [56] Davoodi & Komeili | 2021 | P | G/L | Alg.-agnostic | Heatmaps | × | ✓ |
| [91] Guo et al. | 2021 | P | G | Q-learn./PG | Importance | × | ✓ |
| [149] Madumal et al. | 2020 | P | L | Model-free | Causal Expl. | ✓ | × |
| [148] Madumal et al. | 2020 | P | L | Model-free | Distal Expl. | ✓ | × |
| [99] Herlau & Larsen | 2022 | P | G | Policy | Feature Imp. | × | ~ |
| [79] Gajcin & Dusparic | 2022 | P | L | DQN | Causal relation | × | × |
| [255] Wang et al. | 2023 | P | L | Policy | Causal Expl. | × | ✓ |
| [194] Pynadath et al. | 2018 | P | L | Model-based | Written Expl. | ✓ | × |
| [234] Tabrez et al. | 2019 | P | G | POMDP | Written Expl. | ✓ | ✓ |

(Continued)

**Table 3.1** (Continued)

| Reference | Year | Time | Scope | Agent | Explanation | Human | Automatic |
|---|---|---|---|---|---|---|---|
| [50] Cruz et al. | 2022 | P | L | Not specified | Written Expl. | ✓ | ✓ |
| [87] Greydanus et al. | 2018 | P | G | A3C | Saliency maps | ✓ | ✓ |
| [112] Iyer et al. | 2018 | P | L | DQN | Saliency maps | ✓ | ✓ |
| [92] Gupta et al. | 2019 | P | L | Not Specified | Saliency maps | ✓ | ✓ |
| [107] Huber et al. | 2019 | P | L | Dueling DQN | Saliency maps | × | ✓ |
| [117] Joo & Kim | 2019 | P | L | A3C | Saliency maps | × | ✓ |
| [256] Wang et al. | 2020 | P | L | DQN | SHAP | × | ✓ |
| [200] Rizzo et al. | 2019 | P | L | PG | SHAP | × | ✓ |
| [254] Wang et al. | 2018 | P | L | DQN | Visual anal. | ✓ | ✓ |
| [236] Takagi et al. | 2024 | P | G | State Transitions | Map/Slider | ✓ | ✓ |
| [53] Dao et al. | 2018 | P | G | Double DQN | Snapshots | × | ✓ |
| [165] Mishra et al. | 2018 | P | G | Double DQN | Snapshots | × | ✓ |
| [212] Sequeira & Gervasio | 2020 | P | G | Q-learn. | Short Video | ✓ | ✓ |
| [7] Amir & Amir | 2018 | P | G | Q-values/policy | Importance | × | ✓ |
| [9] Amitai et al. | 2023 | P | G | Policy | Logic statements | ✓ | ✓ |
| [106] Huang et al. | 2018 | P | G | SAC | Importance | ✓ | ~ |
| [140] Liu et al. | 2018 | I | G/L | DQN | Tree/Heatmaps | × | ✓ |
| [220] Silva et al. | 2020 | I | G/L | Q-learn./PG | Tree | ~ | ✓ |
| [48] Coppens et al. | 2019 | P | G/L | PPO | Heatmaps | × | × |
| [245] Vasic et al. | 2022 | I | G | DQN | Tree | × | ✓ |
| [43] Chen et al. | 2022 | P | G/L | Mod.-agnostic | Tree | × | ✓ |
| [151] McCalmon et al. | 2022 | P | G | DQN/PPO | Written Expl. | ✓ | ~ |
| [97] Hein et al. | 2017 | I | G | Model-based | Fuzzy rules | × | ~ |

(Continued)

**Table 3.1** (Continued)

| Reference | Year | Time | Scope | Agent | Explanation | Human | Automatic |
|---|---|---|---|---|---|---|---|
| [98] Hein et al. | 2018 | P | G | Model-based | Boolean/Eq. | × | ✓ |
| [248] Videau et al. | 2022 | I | G | Model-based | Control rules | × | ✓ |
| [51] Custode et al. | 2023 | I | G | Q-learning | Tree | × | ✓ |
| [72] Ferigo et al. | 2023 | I | G | Q-learning | Tree/Heatmaps | × | ✓ |
| [152] McCarthy et al. | 2022 | P | G | Rainbow DQN | Boolean rules | × | ✓ |
| [105] Huang et al. | 2020 | I | G | Model-based | If-Then rule | × | ✓ |
| [247] Verma et al. | 2018 | I | G | Policy | Policy rules | × | ∼ |
| [19] Bastani et al. | 2020 | I | G | Policy | Policy rules | × | ✓ |

Another fundamental point that has to be discussed in depth relates to the usage of a special piece of information: distal information [148]. Specifically, it has been proven that these data help to increase the user's comprehension of the decision taken by the agents, by providing the most salient future moments in the optimal trajectory [148]. However, the determination of which action or state can be labeled as distal has always been manually managed. In this way, the main point of automation no longer applies due to the human intervention in the selection of this information. Moreover, this task can be simplified through the recognition of the significant states. In fact, the actions associated with that occasion can be recognized directly as distal. Although a large variety of importance metrics have been developed in recent times [7, 106, 155, 239], none of the state-of-the-art measures can efficiently recognize all the properties relative to a distal state. Therefore, the following research question is addressed:

> **Research Question 2 (RQ2):** How can the distal state be automatically detected through the adoption of the importance metrics?

Finally, given the methodologies for solving both the research questions proposed until now, it is fundamental to find a way to couple them. The resulting combination can consist of a completely automatic method able to create causal explanations with distal information recovered objectively. The idea of combining the application of human-verified causal models for the production of helpful explanations and importance metrics for the identification of useful data can provide an overall procedure that could be applied to a real-world problem more safely. Hence, the remaining research question regards this interesting mixture:

> **Research Question 3 (RQ3):** How can human-validated causal models and the importance metrics be coupled together to improve the automation of the complete explanation-generator pipeline?

Each of the following chapters focuses their attention on a specific research question. In detail, in the coming Chap. 4, the core topic concerns causality and its automation. Hence, RQ1 is addressed and answered. Subsequently, multiple metrics are introduced to answer RQ2 in Chap. 5. Finally in Chap. 6, the proposed methods are combined and tested in a custom complex environment to verify the last RQ3.

# A Bayesian Network and RNN Methodology for Causal-Distal XRL

**4**

In this chapter, the main focus is on the definition of a methodology that can automatically generate causal explanations for RL agents. Hence, the final goal of this part is to address the RQ1. In the following, the core problem that has to be solved is how to automate the creation of explanations for model-free RL algorithms to answer "Why" and "Why not" questions. Therefore, it is relevant to point out that not only the direct explanations of the action chosen have to be produced but also the counterfactual scenarios have to be verified. To this end, inspired by the proficient work of [148, 149], a combination of multiple techniques regarding both supervised, unsupervised, and reinforcement learning paradigms is adopted. The resulting methodology combines the use of Bayesian Networks for the approximation of the agent and environment interactions and the application of Recurrent Neural Networks for the prediction of the distal information. Then, the output explanations have been evaluated in a user study to compare the improvements in terms of increased trust and comprehension. By comparing explanations obtained with different processes it is possible to derive insights on what humans prefer and recognize as more useful. The procedure and relative outcomes that are reported in this chapter have already been published in [159].

The structure of the chapter is listed below. Before introducing the novel algorithmic approach proposed, the theoretical background is discussed. In detail, after a brief motivation section about the definition of causality and its importance in today's world, the preliminary mathematical concepts employed in the new methodological pipeline are explicated. In this way, the comprehension of the overall theoretical background can be ensured before the discussion of the methods. Subsequently,

---

**Supplementary Information** The online version contains supplementary material available at https://doi.org/10.1007/978-3-658-50495-3_4.


R. Milani, *Advanced Automation for Comprehensible Causal Explanations of Reinforcement Learning Agents*,
https://doi.org/10.1007/978-3-658-50495-3_4

51

the foundational papers of Madumal et al. [148, 149] are analyzed in depth. From the discussion of these articles, originates the proposed workflow to improve the automation using causal models in XRL. At the end, both the computational and user study results are presented.

## 4.1    Preliminaries

In this section, the theoretical background fundamental to ensure the comprehension of the methodologies reported subsequently is presented. The focus lies on the concept of causality and the different mathematical structures that can be employed to model it. After defining the causal models central to this thesis, multiple algorithms able to identify the correct graphical causal representations from data are outlined.

### 4.1.1    Reasoning and Causality

Causality refers to the presence of a relationship between two variables, where the changes in one lead to modifications in the other. Specifically, the former one is defined as *cause* of the second variable, called *effect*. The study of these particular interactions has been pivoted by the work of Judea Pearl [184], who mathematically defined the models and reasoning techniques useful for analyzing causal relations. Solving causal problems requires a wider and deeper knowledge of extensions to standard statistics. While the goal in classical statistical analysis is to determine the parameters of a distribution from samples drawn from that distribution, causal analysis adds another main task: it aims to discover the probabilities under conditions that are changing [185]. Thus, for detecting a causal interaction, a relationship that remains invariant when external variables are modified must be provided. This requirement is also fundamental to delineate the distinction from concepts of association, e.g., correlation, dependence, and likelihood; which can be just defined through distribution functions.

One of the most critical aspects in the field of causality is its distinction from the notion of correlation. To clarify the differences, it is first necessary to provide the definition of correlation: the correlation between two variables indicates the presence of a relationship that can be estimated in strength and direction (in terms of its sign). Nevertheless, this correlation does not prove that if one variable is changing then the other one will modify its values. For this reason, it is crucial to not mistake correlation for causation. While correlation does not imply causality, the opposite is true: causality always implies correlation, meaning that causality is a subset of correlation [184]. A classical example of this question is the following: during a hot summer day, a high number of ice cream were sold and an increase in the occurrence of sunburn cases was recorded. In this case, it is clear how the warm temperature caused the increment in ice cream sales but also the fact of having multiple sunburns. However, although correlated, the variables of ice cream sold and the number of sunburns are not causally dependent. The interactions between these entities can be visualized using the graph shown in Fig. 4.1. With this representation, it is clear how the causality is generated by "higher-level" variables, while correlation appears between "lower-levels". Thereby, it is fundamental to declare mathematically these particular relationships to avoid any possible misunderstanding.

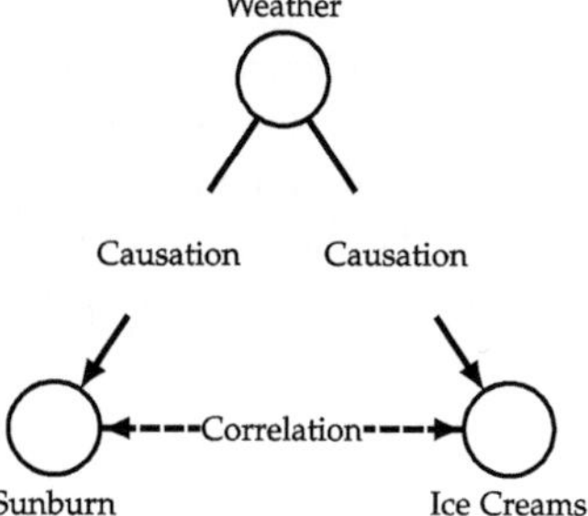


**Fig. 4.1** Example of Causation-Correlation relationship

Before the advent of the following definitions, the common manner of dealing with causal problems involved the mixture of equations and graphs. In particular, the first scientist who attempted to adopt a mathematically sound framework for the recognition of the connections between diseases and symptoms was the geneticist Sewall Wright [264]. In his seminal work, the author associated the variable $X$ to the level of disease and $Y$ to the severity of symptoms. Consequently, he wrote the following system of linear equations:

$$\begin{cases} X = u_X \\ Y = \beta X + u_Y \end{cases}. \tag{4.1}$$

where $u_X$ and $u_Y$ represent all the observed or unobserved background information that are kept unexplained but have a role in modifying $X$ and $Y$, respectively. These variables are called *exogenous* and also include all the errors or disturbances. Besides the exogenous variables, there are other variables called *endogenous* which can be influenced by the exogenous ones. In this specific case, $X$, $Y$ represent the endogenous variables. Lastly, the parameter $\beta \in \mathbb{R}^+$ is defined as the *path coefficient* and represents the causal effect of $X$ on $Y$. Particularly, when $X$ is increasing by a unit, then $Y$ will be incremented by $\beta$ units regardless of the other exogenous variables influences. All this information is fundamental to characterize a *Structural Equation Model* (SEM) [85].

However, with the SEM considered in Equation (4.1), it is not possible to distinguish the directionality of the process. In fact, the second equation can be rewritten as follows:

$$X = \frac{Y - u_Y}{\beta}, \tag{4.2}$$

where, due to the property of algebraic equations of being symmetric, this could provide the misinterpretation that the symptoms influence the disease. Therefore, it is required to extend this mathematical structure with a particular diagram helpful in defining the correct orientation. Given the disease-symptoms SEM in Equation (4.1), the possible causal diagrams that can be employed are shown in Fig. 4.2.

**Fig. 4.2** Examples of associated diagrams to the SEM in Equation (4.1). The exogenous variables are recognized by the dashed lines

In reading these path diagrams, it is important to focus on the structural relationships, such as parent, child, ancestor and descendant. These properties are helpful in the interpretation of which variables are the causes and which are the effect, thus, providing a self-evident explanation of the causality in the specific problem: the parent nodes can be identified as the causes of the changes seen in the children nodes, leading to creating effects in all the descendants as well.

Nonetheless, the selection of a diagram can also lead to diverse outcomes during the statistical analysis. Indeed, looking at the examples displayed in Fig. 4.2a and Fig. 4.2b, it can be observed that, for the former case

$$Cov(X, Y) = \beta \tag{4.3}$$

while, for the latter scenario, due to the correlation between the exogenous variables

$$Cov(X, Y) = \beta + Cov(u_X, u_Y). \tag{4.4}$$

In this example, a major role is played by the possibility of having a correlation between the exogenous variables $u_X$ and $u_Y$. In the presence of correlation, the covariance of the two endogenous variables will not only depend on $\beta$ but also on the aforementioned covariance coefficient. Hence, in this case, the causal relations are encoded not in the links but in the missing connections. While an arrow can identify a probable causal connection, where the strength remains to be determined by the data, a missing edge indicates a claim of zero influence. Similarly, in the case of the missing double arrows representing a zero covariance. Obviously, the representation of the proper path diagrams becomes more challenging with complex situations including a huge number of variables. Despite being a toy problem, the disease-symptoms model studied in [185] illustrates the importance of the selection of the correct diagrams.

Although the SEMs have been developed in the context of linear regressions, their application can be extended to nonlinear and nonparametric models, i.e., when the functional form of the equation is unknown [186]. Here, the invariant property of SEMs can be exploited to derive a special functional form where all the observed variables can be explicated on the left-hand side. Another example can help in the comprehension of this notion. Given the diagram in Fig. 4.3, the corresponding nonparametric interpretation that can be obtained is the following:

$$\begin{cases} Z = f_Z(u_Z) \\ X = f_X(Z, u_X) \\ Y = f_Y(X, u_Y) \end{cases}, \tag{4.5}$$

where $u_X$, $u_Y$, and $u_Z$ can be assumed to be independent from each other. Each of the functions $f_X$, $f_Y$ and $f_Z$ represents a causal process that computes the value of the dependent variable on the left-hand side from the inputs on the right-hand side. It is fundamental to observe that the absence of the variable $Z$ in $f_Y$ arguments assumes that the variations of $Z$ will leave $Y$ unmodified, as long as the other input variables $u_Y$ and $X$ remain constant. Then, a system of functions that are invariant to changes in the form of the other functions is defined as *structural*.

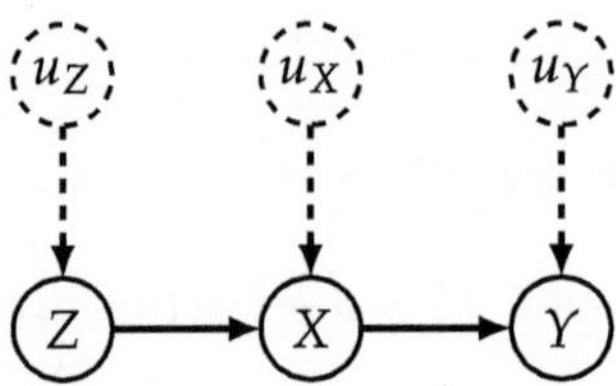

**Fig. 4.3** Diagram associated to the SEM in Equation (4.5)

In [181], Pearl built on these concepts a more advanced notion for the analysis of causal relationships. In particular, the author combined the SEMs, the graphical models for probabilistic reasoning and the potential-outcome framework of Rubin [202] for developing the Structural Causal Models (SCMs). The formal definition is provided below:

**Definition 4.1**  ([190]) *A Structural Causal Model (SCM) $C = (S, P_u)$ consists of a collection S of n structural functions:*

$$X_i = f_i(Pa_i, u_i), \quad i = 1, \ldots, n, \tag{4.6}$$

*where $Pa_i \subseteq \{X_1, \ldots, X_n\} \setminus \{X_i\}$ are the parent variables of $X_i$, and a joint distribution $P_u = P_{u_1,\ldots,u_n}$ over the exogenous variables $u_j$, $j = 1, \ldots, n$, which are required to be jointly independent, i.e., $P_u$ is a product distribution.*

The associated graph $G$ of an SCM is generated by drawing first all the vertices $X_i$ and the edges from each parent node $Pa_i$ to $X_i$, $\forall i = 1, \ldots, n$. This means that for each variable on the left-hand side of the structural equations, there is an incoming arrow from the corresponding endogenous variables on the right-hand side. Furthermore, it is fundamental to assume that a directed acyclic graph (DAG) is produced, i.e., the edges present an orientation and no directed cycles are present.

A simple example is briefly discussed now to clarify the construction of these causal structures, relevant for the full comprehension of this chapter's methodologies. Given the following SCM:

$$\begin{cases} X_1 = f_1(X_2, u_1) \\ X_2 = f_2(X_3, X_4, u_2) \\ X_3 = f_3(u_3) \\ X_4 = f_4(X_3, u_4) \end{cases} \tag{4.7}$$

where $u_1$, $u_2$, $u_3$, $u_4$ are jointly independent. The related graph is shown in Fig. 4.4. Here, the vertex $X_3$ is a parent node for both the variables $X_2$ and $X_4$, since it appears on the right-hand side of both of their respective equations. For the same reason, directed edges from $X_4$ and $X_2$ are respectively going to $X_2$ and $X_1$. Note that in this example, the causal ordering is only one ($3 \rightarrow 4$, $4 \rightarrow 2$, $3 \rightarrow 2$, $2 \rightarrow 1$), hence, the graph is acyclic. Moreover, there is an obvious difference between how the diagrams of the SEMs and SCMs are represented. In the former case, the exogenous variables are included in the plot, while in the latter scenario, only the dependent variables are represented. Despite the SEM graph including more information, the focus will be on the SCMs due to their simpler causal explanation.

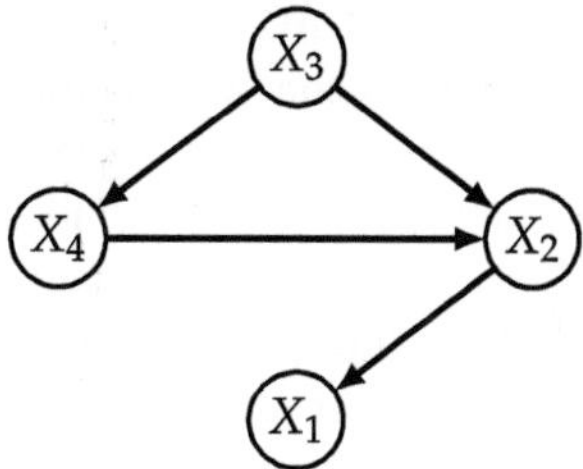

**Fig. 4.4** Diagram associated to the SCM in Equation (4.7)

With this advanced model, it is possible to answer three types of queries: observational, interventional and counterfactual [184]. For the first type, the aim is to determine possible associations between the variables through the answer to questions such as "How does the observation of variable X change the belief of variable Y?". In this way, statistical interactions can be identified. So, the user has to play the role of the observer of the natural phenomenon (passively). In the interventional case instead, it is possible to assume the role of the scientist performing experiments. In detail, one or more variables can be manipulated by fixing them (through the specific do operator) and observing the outcomes of this intervention. By setting its value, a variable is made also independent of its former parent nodes, hence, the causal structure is modified. Thereby, interventional queries are crucial for understanding the potential outcomes of policy changes. Lastly, the counterfactual scenario involves hypothetical assumptions where the parent nodes assume different values. By this gimmick, possible diverse outcomes can be studied, enlarging the overview of the specific problem for the user. For major information regarding these queries, refer to Pearl's book [184].

The ability to explain the causal effects is not the prerogative of only the SCM. A wide range of causal graphs or diagrams can be considered in the solution of this task. In the next section, a different kind of structure is introduced: the *Bayesian Networks*.

## 4.1.2   Bayesian Networks

So far, only causal models based on algebraic structural equations such as SEMs and SCMs have been discussed. Nevertheless, other graphical models can efficiently encode the joint probability distribution and analyze the relationships between variables. To this end, *Bayesian Networks* (BNs) represent a well-suited choice [96].

A BN $B = \{G, P\}$ consists of a DAG $G = (V, E)$ composed of the set of vertices $V$ and set of edges $E$; and of a conditional probability distribution $P$. Each vertex $X_i \in V$, $\forall i = 1, \ldots, n$ represents a variable while the structure defined by the set of the edges indicates the conditional independence assertions of the variables in $V$. Explicitly, the lack of connections between two variables encodes a conditional independence. Regarding the notation used here, the variables are denoted with capital letters, while lowercase letters identify specific values for that variable. Specifically, $P(x_i)$ denotes the probability of the variable $X_i$ to take value $x_i$, i.e., $P(X_i = x_i) = P(x_i)$ [184]. This particular choice helps to not overload the notation, even though, in the necessary cases, it is fully explicated.

The probability distribution $P$ satisfies the following property:

$$P(x) = \prod_{i=1}^{n} P(x_i | Pa_i), \tag{4.8}$$

where, $x = (x_1, \ldots, x_n)$ is a particular realization for each feature $X_i \in V$ and $Pa_i$ are, as in the previous section, the parent nodes of the variable $X_i$ in $G$ [96]. Equation (4.8) involves not only the probabilities but also the structure of the chosen DAG. Moreover, this formula helps to factorize the joint distribution in the product of the conditional probabilities of each variable given its parents' values.

To highlight the importance of Equation (4.8), a minimal example is illustrated. Let $B = \{G, P\}$ be a BN where $G$ is shown in Fig. 4.5. From the chain rule of probability, it holds that

$$P(x) = \prod_{i=1}^{5} P(x_i | x_1, \ldots, x_{i-1}). \tag{4.9}$$

However, under the relationships represented in Fig. 4.5 it is possible to simplify the conditional probabilities as follows:

$$P(x_2 | x_1) = P(x_2), \tag{4.10}$$
$$P(x_3 | x_1, x_2) = P(x_3), \tag{4.11}$$
$$P(x_5 | x_1, x_2, x_3, x_4) = P(x_5 | x_1), \tag{4.12}$$

where the resulting equations are derived from the observation that $x_2$ and $x_3$ do not depend on any other nodes, i.e., the corresponding parent sets are empty, and $x_5$ is influenced only by $x_1$. Hence, the joint probability can be expressed as:

$$P(x) = P(x_1) \, P(x_2) \, P(x_3) \, P(x_4 | x_1, x_2, x_3) \, P(x_5 | x_1). \tag{4.13}$$

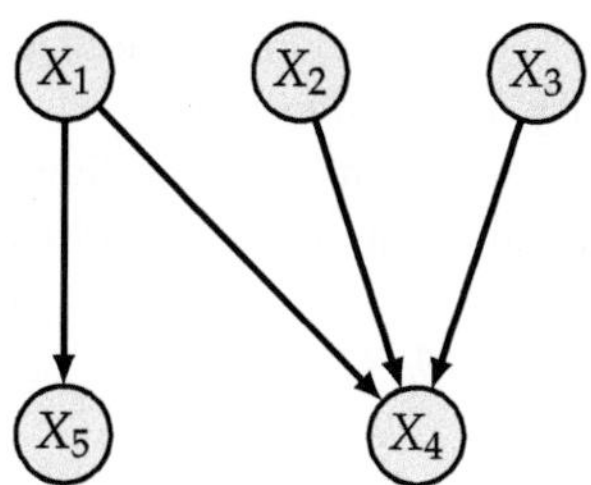


**Fig. 4.5** Example of DAG for a BN

The opposite direction is also possible: if the conditional probabilities are set from the beginning, then the structure of the BN can be derived consequently. The previously explained procedure can be reversed in order to determine the edges in the DAG. However, it is critical to bear in mind that the ordering of the variables plays a relevant role. If this choice is performed carelessly, then the graph obtained could fail to maintain the conditional independencies [96]. Therefore, in the worst-case scenario, $n!$ variable orderings have to be checked to detect the best one. However, there exist other techniques that can discriminate the proper structure for the DAGs in the BN context. This methodology is thoroughly described in Sect. 4.1.3.

The aspect that will be central in the next pages of this section is relative to the computation of a probability of interest, given a model. This procedure is also known as *probabilistic inference* [63]. From now on, the main assumption is that the DAG for the BN has been deduced from data, thus, the focus is only on learning the probabilities involved. Another fundamental requirement is to assume the full observability of the state. In this way, the data gathered can give insights into the full underlying model.

One of the simplest approaches for inferring the probabilities is the frequentist manner. The process relies on the direct computation of the probabilities from the frequencies observed in the data samples. Following the example in Fig. 4.5, if the goal is to evaluate the probability $P(x_4|x_1, x_2, x_3)$, then the data points presenting all their possible values have to be collected in order to estimate the actual probability as follow:

$$P(x_4|x_1, x_2, x_3) = \frac{P(x_1, x_2, x_3, x_4)}{P(x_1, x_2, x_3)} \tag{4.14}$$

$$\approx \frac{|(x_1, x_2, x_3, x_4)|}{|(x_1, x_2, x_3)|}, \tag{4.15}$$

where $|S|$ denotes the number of samples for which the properties $S$ are satisfied. Despite the simplicity of this approach, more advanced techniques have been developed to improve the efficiency of the estimation.

A different methodology is based on the application of the Bayesian method to calculate the local probability distribution for each variable:

$$P(X_i|Pa_i, \theta_i, G), \quad \forall i = 1, \ldots, n, \tag{4.16}$$

where $\theta = (\theta_1, \ldots, \theta_n)$ is the vector of the parameters $\theta_i$, called *Bayesian estimators* [274], for the probability distribution of $X_i$ and $G$ is the DAG of the Bayesian Network. Given the relevance that this approach has in the developed procedure that is described in the following chapters, a detailed description of its pipeline is now outlined.

The Bayesian estimators can be represented in the form of a matrix for each variable $X_i$. In fact, it is possible to decompose them according to different instantiations of the $X_i$ and the variables in the parent set $Pa_i$. By these means, the estimators' components can be denoted as follows:

$$\theta_{ijk} = P(x_i^j | pa_i^k, \theta_i, G) > 0, \tag{4.17}$$

where $i = 1, \ldots, n$, $j = 1, \ldots, r_i$, and $k = 1, \ldots, q_i$, with $r_i$ representing the number of possible values attained by the variable $X_i$ and $q_i$ indicating the number of variables in the parent set $Pa_i$. Also in this case, a reduction of the notation has been made by writing just $x_i^j$ as the meaning of $X_i = x_i^j$ and $pa_i^k$ as the condition $Pa_i = pa_i^k$. In the latter equation, $pa_i^k$ has to be intended a particular vertex in the parent set $Pa_i$. Essential conditions in this framework are that the variables take discrete values and the number of variables is finite. In general, for the aim of this work, the first requirement does not induce any practical problems since it is always possible to discretize a continuous space and analyze the environment in a discrete form. Moreover, as already stated in Chap. 2, the novel methodologies discussed in this thesis are applied in finite MDPs. Thereafter, the number of variables is always finite.

Under these assumptions, let:

$$\theta_i = ((\theta_{ijk})_{j=2}^{r_i})_{k=1}^{q_i}, \quad \forall i = 1, \ldots, n, \tag{4.18}$$

where $\theta_{i1k}$ has been excluded from the list since it can be derived from the other values as $1 - \sum_{j=2}^{r_i} \theta_{ijk}$. For the sake of brevity, the vector of parameters will be denoted as

$$\theta_{ik} = (\theta_{i2k}, \ldots, \theta_{ir_ik}), \quad \forall i = 1, \ldots, n, \ \forall k = 1 \ldots, q_i. \tag{4.19}$$

The last requirement necessary for this estimation to properly work is the fulfilment of the independence condition of the parameter vectors $\theta_{ik}$ [227].

Under these hypotheses, given a random sample $D$ where there are no missing values, the posterior distribution can be factored as follows:

$$P(\theta | D, G) = \prod_{i=1}^{n} \prod_{k=1}^{q_i} P(\theta_{ik} | D, G). \tag{4.20}$$

Therefore, each vector parameter $\theta_{ik}$ can be updated independently as in the one-variable case. Before discussing the calculation of the posterior distribution, it is necessary to introduce the the Dirichlet distribution $Dir(\cdot | \cdot)$ defined through the gamma function $\Gamma(\cdot)$ as:

$$Dir(\theta|\alpha_1, \ldots, \alpha_r) \equiv \frac{\Gamma(\sum_{k=1}^{r} \alpha_k)}{\prod_{k=1}^{r} \Gamma(\alpha_k)} \prod_{k=1}^{r} \theta_k^{\alpha_k - 1}. \tag{4.21}$$

Now, assuming that the prior distribution of each vector $\theta_{ik}$ is $Dir(\theta_{ik}|\alpha_{i1k}, \cdots, \alpha_{ir_ik})$ with parameters $\alpha_{i1k}, \ldots, \alpha_{ir_ik} > 0$, the posterior distribution can be calculated as follows:

$$P(\theta_{ik}|D, G) = Dir(\theta_{ik}|\alpha_{i1k} + N_{i1k}, \ldots, \alpha_{ir_ik} + N_{ir_ik}), \forall i = 1, \ldots, n,$$
$$\forall k = 1, \ldots, q_i \tag{4.22}$$

where $N_{ijk}$ is the number of cases in $D$ such that $X_i = x_i^j$ and $Pa_i = pa_i^k$, and $Dir(\cdot|\cdot)$ is the Dirichlet distribution.

This choice for the prior distribution is justified straightforwardly by the discrete type of variables that are involved in this problem. In fact, this particular situation can be also looked at as rolling a $k$-sided dice $n$ times. Moreover, the Dirichlet distribution, as it will be noticed in the calculations that come, is a conjugate prior to the distribution for categorical and multinomial distributions, i.e., the posterior and the prior belong to the same family of distributions.

To obtain the predictions, it has to be averaged over the possible combinations of $\theta$. For example, if the task is to compute the probability $P(x^{N+1}|D, G)$, where $x^{N+1}$ is the next case seen after $D$, it can be supposed without loss of generality that the instantiation indicates $X_i = x_i^j$, and $Pa_i = pa_i^k$, where obviously $j$ and $k$ depends on $i$. Thus, the following expression needs to be evaluated:

$$P(x^{N+1}|D, G) = \mathbb{E}_{P(\theta|D,G)}\left[\prod_{i=1}^{n} \theta_{ijk}\right]. \tag{4.23}$$

However, to compute this expectation, it is useful to remember that the parameters remain independent given D. Therefore:

$$\mathbb{E}_{P(\theta|D,G)}\left[\prod_{i=1}^{n} \theta_{ijk}\right] = \int \prod_{i=1}^{n} \theta_{ijk} P(\theta|D, G) d\theta \tag{4.24}$$

$$= \prod_{i=1}^{n} \int \theta_{ijk} P(\theta_{ik}|D, G) d\theta_{ik} \tag{4.25}$$

Now, exploiting the following well-known property [96] of the Dirichlet distribution:

$$\int \theta_k \, Dir(\theta|\alpha_1 + N_1, \ldots, \alpha_r + N_r)d\theta = \frac{\alpha_k + N_k}{\alpha + N} \tag{4.26}$$

it results that:

$$P(x^{N+1}|D, G) = \prod_{i=1}^{n} \frac{\alpha_{ijk} + N_{ijk}}{\alpha_{ik} + N_{ik}} \tag{4.27}$$

where $\alpha_{ik} = \sum_{j=1}^{r_i} \alpha_{ijk}$ and $N_{ik} = \sum_{j=1}^{r_i} N_{ijk}$. In this way, the prediction can be directly performed through the adoption of Equation (4.27) for each new datapoint discovered.

The forecasting procedure that has been explained here in the BN context presents a difference from the one considering the structural equations. In fact, in the former scenario, the predictions are principally relying on the statistical dependencies that have been found in the accumulated data. On the other hand, in the SEM and SCM, the changes in the variables are described through pure causal connections defined as algebraic equations. Therefore, there is a critical difference between the causality intended in the BN and the one in the general SCM. The causal relationship discovered by the BNs has to be intended as a structural dependency of the variables. Nevertheless, the encoded dependencies in BN do not always indicate causality effects, due to their probabilistic nature. Hence, the causality considered in BN can be addressed as such in a broader sense [182]. For this reason, the edges of a BN will be recognized as causal connections.

Additionally, another clear difference, that has been already partially highlighted in the previous step, is the presence of an algebraic object in SCMs that is absent in BNs. As long as the causal graphs remain acyclic, the algebraic modification of the SCMs can be interpreted as interventions on the causal systems [184]. Even though this specific operation can not be performed in the BNs, an essentially equivalent symbolic representation of the SEMs can be observed in the Markov factorization, where the causal structure can be reconstructed. Regrettably, this information does not include the proper rate of changes derived from the modification of the variables in the parent set.

Although some differences to SCMs arise, the application of this particular statistical methodology can bring multiple benefits. First, BNs and Bayesian statistical techniques facilitate the integration of the domain knowledge and the data. The causal semantics of BNs make encoding the causal prior knowledge straightforward by adapting the prior distribution used. Moreover, the strength of the causal connections is also weighted by the probabilities inferred. Another interesting opportunity is related to the generalization of the BN to their time-aware version: the *Dynamic Bayesian Networks* (DBNs) [59]. In this context, the interaction between variables in sequential problems can be studied at multiple time steps. For the definition of the DBNs, it is necessary to construct two networks: one related to the usual dependencies between variables at the same time-step and a second transition network illustrating the temporal connections. In this scenario, all the previous ideas applied in BNs can be directly transferred to the DBNs without any loss of generality.

The remaining procedures, that have to be presented in order to focus subsequently on the application of these causal approaches in the XRL context, is the generation of the DAGs from data. This topic is covered in the next section.

## 4.1.3   Learning Causal Structures from Data

In this section, multiple algorithms for learning the causal structures of SEMs and BN from data are introduced. The attention is focused only on the simple linear models of SEMs and not on the SCMs due to the comparison experiments that are carried out in the later stages of the methodological approach proposed. For each of these algorithmic procedures, it is fundamental to characterize the kind of model that is searched. Therefore, it is relevant to highlight the hypothesis under which these processes can be applied.

*LiNGAM*

A first type of model consists of a non-Gaussian variant of SEM and BN called LiNGAM (Linear Non-Gaussian Acyclic Model), firstly presented in [215]. The observed data are assumed to be generated from a process that can be represented by a DAG, having $B \in \mathbb{R}^{n \times n}$ as its transposed adjacency matrix. In detail, $B_{ij}$ indicates the connection strength from the variable $x_j$ to $x_i$. Moreover, $k(i)$ denotes the causal order of the variables $x_i, \forall i = 1, \ldots, n$ such that no successive variables determine or have a direct path to any precedent variables. Furthermore, the model assumes that the relations are linear, so it can be mathematically defined as follows:

$$x_i = \sum_{k(j)<k(i)} B_{ij}x_j + e_i, \quad \forall i = 1, \ldots, n, \tag{4.28}$$

where $e_i \in \mathbb{R}^n$ are the external influences recognized as continuous random variables, each having a non-Gaussian distribution with zero mean and non-zero variance. Furthermore, they are also independent to ensure that there are no latent confounding variables [228]. Compactly, the equation above can be rewritten in matrix form as follows:

$$x = Bx + e \tag{4.29}$$

where $x$ is the random (variables) vector, and $B$ could be permuted simultaneously for the row and column to be strictly lower triangular, in order to satisfy the acyclicity condition. By algebraically solving Equation (4.29) for $x$, it is obtained the following expression:

$$\begin{aligned} x &= (I - B)^{-1}e \\ &= Ae \end{aligned} \tag{4.30}$$

where $A = (I - B)^{-1}$. Now, given the assumptions of $e$ being independent and non-Gaussian, Equation (4.30) represents the independent component analysis (ICA) model [110], which can be efficiently solved through an ICA algorithm, e.g., with FastICA [109]. In this way, $A$ and its inverse $A^{-1}$ can be estimated. Let $W = A^{-1}$, it is interesting to remark that this inverse is close to the final solution needed: $W = I - B$. However, the solution found by the ICA algorithms has permutation, scaling and sign indeterminacies. Indeed, ICA algorithms actually determine $W_{ICA} = PDW$, where $P$ is an unknown permutation matrix and $D$ is an unknown diagonal matrix. Nevertheless, in LiNGAM it can be found the proper permutation matrix $P$ [215] such that the resulting matrix $\tilde{W} = PW$ does not have zeros in the diagonal. To correct the scaling factor of the application of $D$, each row of $\tilde{W}$ can be divided by the diagonal element so that in the diagonal there are only 1s. Finally, can be determined the correct order such that the final matrix is lower triangular. To this end, due also to the numerical approximation caused by the computational algorithm adopted, an iterative approach eliminating the smallest (in absolute value) element is performed to achieve the lower triangular structure. Since the central idea of this approach is based on the ICA models, the overall algorithm is called ICA-LiNGAM algorithm. The full procedure is reported in Algorithm 4.1.

---

**Algorithm 4.1** ICA-LiNGAM algorithm [215]

---

**Require:** Random (variables) vector $x \in \mathbb{R}^n$, its observed Data Matrix $X \in \mathbb{R}^{n \times m}$, threshold lower triangular form $\varepsilon > 0$.

1: Apply an ICA algorithm to obtain an estimate of $A$ s.t.

$$x = Ae$$

2: Find unique permutation $P$ of rows of $W = A^{-1}$ s.t. the resulting matrix $\tilde{W}$ satisfies:

$$\tilde{W}_{ii} \neq 0 \quad \forall i = 1, \ldots, n$$

3: Create a new rescaled matrix $\tilde{W}'$ from $\tilde{W}$ using the following update:

$$\tilde{W}'_{ij} \leftarrow \frac{\tilde{W}_{ij}}{\tilde{W}_{ii}} \quad \forall i, j = 1, \ldots, n$$

4: Compute an estimate of the adjacency matrix $\hat{B}$ solving the equation:

$$\hat{B} = I - \tilde{W}'$$

5: Find the permutation matrix $\tilde{P}$ such that $\tilde{B} = \tilde{P}\hat{B}\tilde{P}^T$ is as approximately a lower triangular matrix so:

$$\sum_{i<j} \tilde{B}_{ij}^2 \leq \varepsilon \quad \forall i = 2, \ldots, n,$$

6: **if** $\sum_{i<j} \tilde{B}_{ij}^2 \leq \varepsilon \quad \forall i = 2, \ldots, n$, **then**

7:    **return** $\tilde{B}$

8: **else**

9:    Set $\frac{n(n+1)}{2}$ smallest (in absolute value) element of $\hat{B}$ equal to 0.

10:    **while** $\exists i$ s.t. $\sum_{i<j} \tilde{B}_{ij}^2 > \varepsilon$ **do**

11:      Find the permutation matrix $\tilde{P}$ such that $\tilde{B} = \tilde{P}\hat{B}\tilde{P}^T$ is as approximately a lower triangular matrix

12:      **if** $\exists i$ s.t. $\sum_{i<j} \tilde{B}_{ij}^2 > \varepsilon$ **then**

13:        Set the next smallest (in absolute value) element of $\hat{B}$ equal to 0

14:      **end if**

15:    **end while**

16:    **return** $\tilde{B}$

17: **end if**

---

Nevertheless, this methodology presents several problems. The first one concerns the adoption of the ICA algorithms. It has been shown that, under some specifically bad conditions on the choices of the initial guess, these approaches may not converge to the solution in a finite number of steps [101]. Additionally, the permutations applied are not scale-invariant. As a consequence, they could provide the wrong ordering of the variables depending on scales and standard deviations of the variables. Therefore, an advanced technique to estimate the causal ordering and the connection strengths was developed by the same authors: DirectLiNGAM [216].

*DirectLiNGAM*

In DirectLiNGAM, the main idea is to first search the correct order for the variables in order to avoid the application of permutations. To determine this rank, it is first fundamental to perform a least square regression of $x_i$ using $x_j$, $\forall i \in U \setminus K$, and $i \neq j$, where $U$ is the set of the subscripts of the variables $x_j$, i.e., $U = \{1, \ldots, n\}$, and $K$ is the set of ordered variables already eliminated. This leads to the identification of the residuals that can be written as

$$r_i^{(j)} = x_i - \frac{cov(x_i, x_j)}{var(x_j)} x_j, \quad \forall i \in U \setminus K, \forall j \in U \setminus K, \text{ and } i \neq j, \qquad (4.31)$$

where $cov(x_i, x_j)$ and $var(x_i)$ denote respectively the covariance between $x_i$ and $x_j$ and the variance of variable $x_i$. The residuals play an essential role in the recognition of the correct order since they show whether a variable is exogenous. In fact, if $x_j$ is independent of its residuals $r_i^{(j)}$, $\forall i \neq j$, then the variable $x_j$ is exogenous (Lemma 1 in [216]). It may be interesting to notice that also the opposite direction of the implication is verified. Hence, it is now critical to identify a measure of independence. The one suggested from the original paper proposing this novel methodology is the mutual information $MI(y_1, y_2)$ [110]. Bach and Jordan [14] derived a nonparametric estimator of the mutual information through the use of the kernel methods and a set of $m$ observations of $y_1$, $y_2$. In particular, they considered the Gaussian kernel values $K_1(y_1^{(i)}, y_1^{(j)})$ and $K_2(y_2^{(i)}, y_2^{(j)})$ (with $i, j = 1, \ldots, m$), computed as follows:

$$K_1(y_1^{(i)}, y_1^{(j)}) = exp(-\frac{1}{2\sigma^2} \|y_1^{(i)} - y_1^{(j)}\|^2), \qquad (4.32)$$

$$K_2(y_2^{(i)}, y_2^{(j)}) = exp(-\frac{1}{2\sigma^2} \|y_2^{(i)} - y_2^{(j)}\|^2), \qquad (4.33)$$

where $\sigma > 0$ is the bandwidth of Gaussian kernel (the suggested value is $\sigma = 0.5$ for $m < 1000$ [14]). Let $K_1$ and $K_2$ denote the corresponding Gram matrices of the aforementioned Gaussian kernels and let $k$ be a small positive constant (the suggested value is $k = 2 \times 10^{-3}$ for $m < 1000$ [14]). The estimator of mutual information is defined as:

$$MI_{kernel}(y_1, y_2) = -\frac{1}{2} \log \frac{\det \mathcal{K}_k}{\det \mathcal{D}_k}, \tag{4.34}$$

where

$$\mathcal{K}_k = \begin{bmatrix} (K_1 + \frac{mk}{2}I)^2 & K_1 K_2 \\ K_2 K_1 & (K_2 + \frac{mk}{2}I)^2 \end{bmatrix}, \tag{4.35}$$

$$\mathcal{D}_k = \begin{bmatrix} (K_1 + \frac{mk}{2}I)^2 & 0 \\ 0 & (K_2 + \frac{mk}{2}I)^2 \end{bmatrix}. \tag{4.36}$$

Coming back to the problem of determining the independence between $x_j$ and the residuals, it is possible to evaluate the independence using the following function:

$$T_{kernel}(x_j; U) = \sum_{i \in U, i \neq j} MI_{kernel}(x_j, r_i^{(j)}). \tag{4.37}$$

In this way, the variable that attains the minimum value for $T_{kernel}$ will be the least causally important. Furthermore, this process has been proved to always find the next exogenous variable if iterated over the reduced set of subscripts, i.e., after eliminating the least relevant feature from $U$ (Lemma 2 in [216]).

After detecting the order, it remains to use this information for generating the strictly lower triangular adjacency matrix $B$. This final step can be completed through the estimation of the connection strengths $B_{ij}$ by using a covariance-based regression, e.g., least squares and maximum likelihood approaches on the observed data matrix $X$ and vector $x$. The full DirectLiNGAM algorithm is summarized in Algorithm 4.2.

---

**Algorithm 4.2** DirectLiNGAM algorithm [216]

---

**Require:** Random (variables) vector $x \in \mathbb{R}^n$, its variable subscripts $U$ and its observed Data
   Matrix $X \in \mathbb{R}^{n \times m}$, Ordered list of variables $K = \emptyset$.
1: **while** $|K| \neq n - 1$ **do**
2:     Perform least squares regression of $x_i$ using $x_j$, $\forall i \in U \setminus K$, and $i \neq j$.
3:     Compute the residual matrix $R^{(j)}$ from $X$ and the residuals $r^{(j)}$ as follows:

$$r_i^{(j)} = x_i - \frac{cov(x_i, x_j)}{var(x_j)} x_j, \quad \forall i \in U \setminus K, \forall j \in U \setminus K, \text{ and } i \neq j,$$

4:     Find $x_l$ s.t.

$$x_l = \arg \min_{j \in U \setminus K} T_{kernel}(x_j; U \setminus K),$$

5:     Append $l$ at the end of $K$.
6:     Set $x = r^{(l)}$, $X = R^{(l)}$.
7: **end while**
8: Append the last remaining variable at the end of $K$.
9: Construct a strictly lower triangular matrix $B$ following the order in $K$ using least squares
   and maximum likelihood approaches applied to $x$ and $X$.
10: **return** $B$

---

While DirectLiNGAM provides better guarantees in terms of finding the optimal
solution in a finite amount of iterations, it can not be extended easily to the time
series cases. Nonetheless, one of the tasks of this chapter is to discover the causal
relationship in the sequential problems (MDPs). Thus, approaches that can take into
account multiple time steps may be fitting to this scenario. For these reasons, next
is introduced a combination of the basic LiNGAM approach with the classic vector
autoregressive models (VAR). So, the resulting method is called VARLiNGAM
[111].

*VARLiNGAM*

To better understand the algorithm, it is necessary first to describe the model chosen.
Let $x_i(t)$, $i = 1, \ldots, n$, $t = 1, \ldots, T$ be the observed time series, where $i$ is the
usual component index and $t$ represents the time index. Hereby, it can be compactly
defined as the vector $x(t)$. From the combination of autoregressive modelling and
SEM, the following model is derived:

$$x(t) = \sum_{\tau=0}^{t_d} B_\tau x(t - \tau) + e(t), \tag{4.38}$$

where, $t_d$ indicates the order of the autoregressive model, i.e., the number of time delays used, $B_\tau \in \mathbb{R}^{n \times n}$ are the transpose adjacency matrices as in LiNGAM for $\tau = 0, \ldots, t_d$, and $e_t$ are the random processes associated to the disturbances. As for the LiNGAM case, in order to ensure the existence of the solution, it is necessary to assume that $e_i(t)$ are mutually independent $\forall i = 1, \ldots, n$ and $\forall t = 1, \ldots, T$ and that they are non-Gaussian. It is interesting to note that, in the case $\tau = 0$, the matrix $B_0$ represents the same acyclic graph as in the causal analysis made for LiNGAM. Therefore, it is possible to exploit the already explained LiNGAM algorithm to find this first information.

Now, from the definition of the model in Equation (4.38), it holds:

$$x(t) = \sum_{\tau=0}^{t_d} B_\tau x(t - \tau) + e(t) \iff \tag{4.39}$$

$$x(t) = B_0 x(t) + \sum_{\tau=1}^{t_d} B_\tau x(t - \tau) + e(t) \iff \tag{4.40}$$

$$(I - B_0)x(t) = \sum_{\tau=1}^{t_d} B_\tau x(t - \tau) + e(t) \iff \tag{4.41}$$

$$(I - B_0)\left[ x(t) - \sum_{\tau=1}^{t_d} (I - B_0)^{-1} B_\tau x(t - \tau) \right] = e(t) \iff \tag{4.42}$$

$$x(t) - \sum_{\tau=1}^{t_d} (I - B_0)^{-1} B_\tau x(t - \tau) = B_0 \left[ x(t) - \sum_{\tau=1}^{t_d} (I - B_0)^{-1} B_\tau x(t - \tau) \right] + e(t) \tag{4.43}$$

where Equation (4.43) represents a LiNGAM model with the residuals values $x(t) - \sum_{\tau=1}^{t_d} (I - B_0)^{-1} B_\tau x(t - \tau), \forall t = 1, \ldots, T$.

Let now $M_\tau = (I - B_0)^{-1} B_\tau$ and $n(t) = (I - B_0)^{-1} e(t)$. So, from Equation (4.41), it is possible to isolate the variable $x(t)$ as follows:

$$x(t) = \sum_{\tau=1}^{t_d} (I - B_0)^{-1} B_\tau x(t - \tau) + (I - B_0)^{-1} e(t) \tag{4.44}$$

$$= \sum_{\tau=1}^{t_d} M_\tau x(t - \tau) + n(t), \tag{4.45}$$

where in the second equality, the values have been substituted for $M_\tau$ and $n(t)$. In particular, it is possible to show that:

$$n(t) = B_0 n(t) + e(t), \tag{4.46}$$

implying that the residuals $n(t)$, which can be also rewritten as $n(t) = x(t) - \sum_{\tau=1}^{t_d} (I - B_0)^{-1} B_\tau x(t - \tau)$ from Equation (4.42), can be represented using the LiNGAM model.

Therefore, the main idea of this approach is based on first estimating $\tilde{M}_\tau$ using a least square regression considering the model in Equation (4.45). After this, the residuals $\tilde{n}(t)$ are approximated by subtracting the real values and the regression values found in the previous step, i.e., $\tilde{n}(t) = x(t) - \sum_{\tau=1}^{t_d} \tilde{M}_\tau x(t - \tau)$. From these values, the ICA-LiNGAM algorithm is applied to estimate the value of $\tilde{B}_0$, obtained from the model in Equation (4.46). Finally, the (approximated) adjacency matrices $\tilde{B}_\tau$ are computed from the definition of $M_\tau$ by multiplying between them all the previously determined values. The full algorithm is shown in Algorithm 4.3.

---

**Algorithm 4.3** VARLiNGAM algorithm [111]

---

**Require:** Random (variables) vector $x(t) \in \mathbb{R}^n, \forall t = 1, \ldots T$, its observed Data Matrix $X(t) \in \mathbb{R}^{n \times m}, \forall t = 1, \ldots, T$, threshold lower triangular matrix $\varepsilon > 0$.

1: Estimate $\tilde{M}_\tau$ using the classical autoregressive model:

$$x(t) = \sum_{\tau=1}^{t_d} \tilde{M}_\tau x(t - \tau) + \tilde{n}(t), \quad \forall t = 1, \ldots, T,$$

2: Compute the residuals $\tilde{n}(t), \forall t = 1, \ldots, T$ as follows:

$$\tilde{n}(t) = x(t) - \sum_{\tau=1}^{t_d} \tilde{M}_\tau x(t - \tau), \quad \forall t = 1, \ldots, T,$$

3: Apply ICA-LiNGAM (Algorithm 4.1) to the following model:

$$\tilde{n}(t) = \tilde{B}_0 \tilde{n}(t) + e(t), \quad \forall t = 1, \ldots, T,$$

and estimate $\tilde{B}_0$.

4: Compute $\tilde{B}_\tau$ from $\tilde{M}_\tau$ and $\tilde{B}_0$ as follows:

$$\tilde{B}_\tau = (I - \tilde{B}_0) \tilde{M}_\tau, \quad \forall \tau > 0,$$

5: **return** $\tilde{B}_\tau, \forall \tau > 0$.

---

*NOTEARS*

A different way of tackling the problem of discovering the causal structures encoded as DAGs is to describe it as an optimization problem. This situation can indeed be modelled as follows:

$$\min_{B \in \mathbb{R}^{n \times n}} F(B)$$
$$\text{subject to } G(B) \in DAGs, \tag{4.47}$$

with

$$F(B) = l(B; X) + \lambda \|B\|_1 = \frac{1}{2n} \|X - XB\|_F^2 + \lambda \|B\|_1, \tag{4.48}$$

$$DAGs = \text{Set of DAGs} \tag{4.49}$$

where $G(B)$ denotes the graph generated from the adjacency matrix $B \in \mathbb{R}^{n \times n}$, $X \in \mathbb{R}^{m \times n}$ represents the observed data, $\lambda$ is the regularization coefficient, $\|\cdot\|_F$ denotes the Froebenius norm and $\|\cdot\|_1$ the $l_1$−norm. The aim is to find the adjacency matrix that could simultaneously approximate in the best way the data following the usual linear SEM as in the previous methods and maintain a sparse structure. For these reasons, the objective function is composed of a first component indicating the loss of the SEM estimations, i.e., $l(B; X) = \frac{1}{2n} \|X - XB\|_F^2$, and the second one for reducing the number of connections, i.e., $\lambda \|B\|_1$.

Despite this optimization problem being an **NP**-complete [45] combinatorial problem, it can be reformulated as a continuous problem. To this end, it is fundamental to rewrite the combinatorial constraint characterizing the DAGs as a continuous condition. This task is accomplished by employing the function:

$$h(B) = tr(e^{B \circ B}) - n, \tag{4.50}$$

where $tr(B)$ is trace of the matrix $B$ and $\circ$ denotes the Hadamard product. The function $h : \mathbb{R}^{n \times n} \to \mathbb{R}$ is a smooth function counting the number of cycles present in the graph generated by an adjacency matrix. Hence, the level set at zero characterizes the directed acyclic graphs (Theorem 1 in [272]).

The resulting continuous optimization problem is:

$$\min_{B \in \mathbb{R}^{n \times n}} F(B)$$
$$\text{subject to } h(B) = 0. \tag{4.51}$$

To solve this problem, the augmented Lagrangian method can be applied, which resolves the problem in Equation (4.51) by augmenting its objective function by a quadratic penalty value depending on the constraint condition. Thus, the problem can be rewritten as:

$$\min_{B \in \mathbb{R}^{n \times n}} F(B) + \frac{\rho}{2}|h(B)|^2$$
$$\text{subject to} \qquad h(B) = 0, \tag{4.52}$$

where $\rho > 0$ is a constant for the penalty derived from not satisfying the constraint. The next step is to find the dual problem that can be obtained from writing the Lagrangian:

$$L^\rho(B, \alpha) = F(B) + \frac{\rho}{2}|h(B)|^2 + \alpha h(B), \tag{4.53}$$

where $\alpha \in \mathbb{R}$ is the Lagrangian multiplier and $L^\rho(B, \alpha)$ is the augmented Lagrangian function. The addition of the quadratic term $\frac{\rho}{2}|h(B)|^2$ to the classical Lagrangian improves the convergence of the duality algorithms described below [74]. Moreover, thanks to the presence of the term $h(B)$, the exact solution can be determined without requiring that $\rho$ tend to infinity, unlike the usual penalization methods [74].

From this, the dual function can be defined as follows:

$$D(\alpha) = \min_{B \in \mathbb{R}^{n \times n}} L^\rho(B, \alpha), \tag{4.54}$$

which is fundamental for the definition of the dual problem:

$$\max_{\alpha \in \mathbb{R}} D(\alpha). \tag{4.55}$$

The final solution to the initial problem in Equation (4.52) can be found in three main steps. First solve the primal problem and find $D(\alpha)$ as in Equation (4.54) and identify $B^*_\alpha$ as the local minimizer depending on the Lagrange multiplier $\alpha$. Then, the dual problem in Equation (4.55) can be solved through the classical dual ascent method. In particular, for this scenario, the dual function is linear for $\alpha$, so the gradient is $\nabla D(\alpha) = h(b^*_\alpha)$. From this calculation, the updating rule following the gradient ascent method is:

$$\alpha \leftarrow \alpha + \rho h(B^*_\alpha), \tag{4.56}$$

where the choice of the stepsize fixed to $\rho$ was decided from the convergence results proved in the Corollary 11.2.1 in [176]. Despite the absence of studies in the worst-case scenario, in practical situations, the complexity of finding a solution does not require many iterations ($t \sim 10$). The repetition in solving first the primal and then the dual problems is stopped in the case when the solution found achieves the constraint condition, i.e., $h(B^*) < \varepsilon_M$, where $\varepsilon_M > 0$ is a positive threshold. Finally, the adjacency matrix discovered in this procedure is modified in the last phase of thresholding: all the edges with weights lower than a threshold parameter $\varepsilon > 0$ are eliminated. In this way, the possible false discoveries are cancelled. The overall process is summarized in Algorithm 4.4.

---

**Algorithm 4.4** NOTEARS algorithm [272]

---

**Require:** Initial guess $(B_0, \alpha_0)$, observed data matrix $X \in \mathbb{R}^{m \times n}$, progress rate $c \in (0, 1)$, tolerance $\varepsilon_M > 0$, threshold $\varepsilon > 0$.

1: **for** $t = 0, 1, \cdots, \infty :$ **do**

2:    Solve the primal problem:

$$B_{t+1} \leftarrow \arg \min_{B \in \mathbb{R}^{n \times n}} (L^\rho(B, \alpha_t)) \text{ with } \rho \text{ s.t. } h(B_{t+1}) < c \cdot h(B_t).$$

3:    Dual ascent:

$$\alpha_{t+1} \leftarrow \alpha_t + \rho h(B_{t+1})$$

4:    **if** $h(B_{t+1}) < \varepsilon_B$ **then**

5:        Store the found solution

$$B^*_{approx} \leftarrow B_{t+1}$$

6:        **break**

7:    **end if**

8: **end for**

9: Thresholding:

$$B^* \leftarrow B^*_{approx} \circ 1_{[|B^*_{approx}| > \varepsilon]}$$

10: **return** $B^*$

---

*DYNOTEARS*

As previously described for LiNGAM, the NOTEARS algorithm can also be extended to consider time-dependent random variables using the classical autoregressive model:

$$x(t) = Bx(t) + \sum_{\tau=1}^{t_d} B_\tau x(t - \tau) + e(t) \tag{4.57}$$

where all the definitions are as in Equation (4.38). The only modification in the notation is the adoption of $B$ instead of $B_0$. In this manner, it is possible to maintain a notation more similar to the problem formulated in NOTEARS. From the single observation, the general matrix form can be generated as follows:

$$X = XB + \sum_{\tau=1}^{t_d} Y_\tau B_\tau + E, \tag{4.58}$$

where $X \in \mathbb{R}^{m \times n}$ is the observed data matrix, $Y_1, \ldots, Y_{t_d} \in \mathbb{R}^{m \times n}$ are the time lagged versions of $X$ and $E \in \mathbb{R}^{m \times n}$ is the matrix of the disturbances. Moreover, let the time-lagged matrices be written as $Y = [Y_1 | \ldots | Y_{t_d}] \in \mathbb{R}^{m \times t_d n}$ and $A = [B_1^T | \ldots | B_{t_d}^T]^T \in \mathbb{R}^{t_d n \times n}$. With this compact notation, the SEM in Equation (4.57) assumes the following form:

$$X = XB + YA + E. \tag{4.59}$$

From this, it is possible to derive the same continuous optimization problem by repeating the same steps performed in the NOTEARS case. The problem that is analyzed in the time-lagged version is the following:

$$\min_{B \in \mathbb{R}^{n \times n}, A \in \mathbb{R}^{t_d n \times n}} F(B, A)$$
$$\text{subject to } h(B) = 0. \tag{4.60}$$

where the modifications from Equation (4.51) lie in the formulation of the loss function for the SEM considering Equation (4.59), and in the addition of a regularization factor for the time-lagged matrices used for increasing the sparsity of the graph, i.e., $\|A\|_1$. The objective function has the following form:

$$F(B, A) = l(B, A) + \lambda_B \|B\|_1 + \lambda_A \|A\|_1, \tag{4.61}$$

where $\lambda_A, \lambda_B > 0$ are the regularization parameters and $l(B, A)$ is the SEM loss function:

$$l(B, A) = \|X - XB - YA\|_F^2. \tag{4.62}$$

It is straightforward to apply the augmented Lagrangian method and solve it as done previously. The augmented Lagrangian function can be written as:

$$L^\rho(B, A, \alpha) = F(B, A) + \frac{\rho}{2}|h(B)|^2 + \alpha h(B), \tag{4.63}$$

and consequently, the dual function is:

$$D(\alpha) = \min_{B \in \mathbb{R}^{n \times n}, A \in \mathbb{R}^{td^n \times n}} L^\rho(B, A, \alpha). \tag{4.64}$$

Finally, the dual problem can be derived:

$$\max_{\alpha \in \mathbb{R}} D(\alpha). \tag{4.65}$$

Given the similarities to the NOTEARS algorithm, the pipeline now deployed for solving the optimization problem in Equation (4.65) is the same: first solve the primal problem in Equation (4.64) finding the optimal values $B_\alpha^*$, $A_\alpha^*$ that are then plugged into the dual problem. Hence, the gradient ascent method is applied to update the Lagrange multiplier. These steps are repeated until the condition on the acyclicity is satisfied, up to a threshold value $\varepsilon_B > 0$. Finally, the thresholding phase is also reported to correct not only the connection at the next time step in $B^*$ but also in the inter-slices $A^*$. Since this is a dynamic version of the NOTEARS algorithm, the authors in [179] called this procedure DYNOTEARS. The complete algorithm is summarized in Algorithm 4.5.

All the approaches presented in this section will have high importance in the proposed methodology. However, before describing the improvements brought in the generation of causal explanations for XRL, it is necessary to analyze the specific literature used as starting point of this research.

---

**Algorithm 4.5** DYNOTEARS algorithm [179]

---

**Require:** Initial guess $(B_0, A_0, \alpha_0)$, observed data matrix $X \in \mathbb{R}^{m \times n}$, and time-lagged version $Y \in \mathbb{R}^{m \times t_d n}$, progress rate $c \in (0, 1)$, tolerance $\varepsilon_M > 0$, thresholds $\varepsilon_B, \varepsilon_A > 0$.

1: **for** $t = 0, 1, \cdots, \infty :$ **do**
2:     Solve the primal problem:

$$B_{t+1}, A_{t+1} \leftarrow \arg \min_{B \in \mathbb{R}^{n \times n}, A \in \mathbb{R}^{t_d n \times n}} (L^{\rho}(B, A, \alpha_t))$$

with $\rho$ s.t. $h(B_{t+1}) < c \cdot h(B_t)$.

3:     Dual ascent:

$$\alpha_{t+1} \leftarrow \alpha_t + \rho h(B_{t+1})$$

4:     **if** $h(B_{t+1}) < \varepsilon_B$ **then**
5:         Store the found solution

$$B^*_{approx} \leftarrow B_{t+1}$$
$$A^*_{approx} \leftarrow A_{t+1}$$

6:         **break**
7:     **end if**
8: **end for**
9: Thresholding:

$$B^* \leftarrow B^*_{approx} \circ 1_{[|B^*_{approx}| > \varepsilon_B]}$$
$$A^* \leftarrow A^*_{approx} \circ 1_{[|A^*_{approx}| > \varepsilon_A]}$$

10: **return** $B^*, A^*$

---

## 4.2  Specific Literature: An Action Influence Model Approach

In this section, an in-depth analysis of the papers by Madumal et al. [148, 149] is provided. These works characterize the field of XRL since chronologically they were the first to apply causal reasoning for the generation of explanations for RL agents. Specifically, the authors assumed that the RL algorithm involved in their experiments covered the subgroup relative to the model-free approaches. Moreover, the environments where the tests were carried out can be represented as finite MDPs. Both the cases of continuous and discrete scenarios were taken into account during the experimentation. The main task that has to be solved is the creation of explanations for "Why Action A?" and "Why not Action B?" questions. The statement produced has to not only be accurate enough to predict the real decisions of the agent and the corresponding responses of the environment but also be understandable to increase the trust of the human users in those RL policies.

In the following, first, the mathematical formulation of the explanations is given, and then, it is clarified how to practically generate them in a specific benchmark. In this way, the comprehension of the action influence model approach for XRL can be extensively improved. Finally, after the overall analysis, a brief description of the major issues related to this methodology is outlined.

## 4.2.1   Mathematical Formulation and Practical Examples

In order to accomplish the goal of answering "Why" and "Why not" questions, Madumal et al. [149], utilized a particular causal graph for the creation of *minimally complete explanations*: the *action influence model*. Before introducing the mathematical definition, it is helpful to grasp the main concept behind this particular causal structure. An action influence graph is composed of a set of nodes representing the features of the states in the MDPs addressed and a set of edges indicating the specific action selected. Therefore, a connection between two vertices in the action influence model denotes how the node in the parent set is influencing the child node through that action. So, the general structure of the action influence models is the same as in the SCMs (Definition 4.1) except that each edge is associated with an action. Formally, an action influence model is defined as follows:

**Definition 4.2**   (Reformulated from [149]) *An action influence model is a tuple* $\mathcal{M} = (S, S_a)$, *where* $S$ *is the set of the structural equations (as in Definition 4.1) and* $S_a = (\mathcal{U}, \mathcal{X}, \mathbb{A})$ *is a signature including the set of exogenous variables* $\mathcal{U}$, *the set of endogenous variables* $\mathcal{X}$, *and the set of actions of a MDP* $\mathbb{A}$. *A structural function* $f_{X_i, a_j} \in S$ *for* $a_j \in \mathbb{A}$, *defines the causal effect on* $X_i$ *from applying action* $a_j$.

From the definition above, it is helpful to declare a class in the endogenous variables that will be later necessary for the production of the explanations. The set of the *reward variables* $\mathcal{X}_r \subseteq \mathcal{X}$ is the set of nodes with an out-degree of 0. This particular group of nodes can be quickly recognized in the action influence graph since they are all sink nodes, i.e., there are only incoming edges but no outgoing connections.

Due to the complexity of the theoretical definition above, it can be beneficial to apply this concept in a practical example. To this end, along this section, the case of creating explanations for agents involved in solving the Taxi environment is examined (for more information about this problem, refer to Appendix A.1 in the

Electronic Supplementary Material). In this way, a clarification of how the formal definitions are applied can be presented.

First, it is fundamental to correctly identify an action influence model for the aforementioned MDP. This particular process can be executed by an expert who is able to recognize all the major aspects and insights of a specific problem. The action influence graph taking into consideration the trivial causal effects in the Taxi environment is reported in Fig. 4.6. Concretely, the variables involved in the graph (taxi row and column position, passenger position and destination) are only endogenous variables since the Taxi environment is a fully-observable environment.

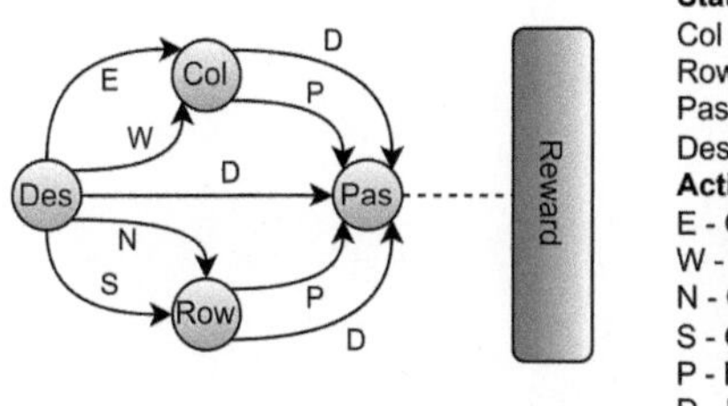


**Fig. 4.6** Action influence graph derived from the Taxi environment

More complicated is the situation for the discrimination of the links between these nodes. The connections between the destination node and the taxi position features are labelled according to the specific movements changing the latter variables. In fact, it is clear how the destination of the passenger is influencing the movements inside of the map. For the passenger node, the actions referring to picking up the passenger and dropping them off are associated with the edges coming from the row and column taxi location and going to the passenger feature. This particular choice is derived from the circumstances in which the passenger can enter the taxi, i.e., when the car is at the same location as the client. Additionally, the destination also influences the passenger position due to the condition of the client being transported to the correct location. Thus, these two nodes are connected through the drop-off operation.

Finally, the features that are strictly related to the achievement of the goal, which are the passenger and destination position, are linked to the reward node. Despite this condition, the only variable that will be indicated as a reward variable is the passenger position, since it is the only sink node between the two. With this particular causal structure, the explanations can be generated using the following definitions.

Before jumping to the definition of the explanations, the introduction of another notion is necessary: an *actual instantiation* of a model $\mathcal{M}$ is indicated as $\mathcal{M}_{\mathcal{X} \leftarrow x}$ in which $x$ is the vector of state variable values from an MDP. In this manner, a precise set of values of all the state variables for the model is selected. Practically in the Taxi example, suppose that the actual state features are as in Fig. 4.7, where the customer is in the "South-West" position, the destination is in "South-East", and the car is in "North" and "East". These data illustrate the actual instantiation of the environment.

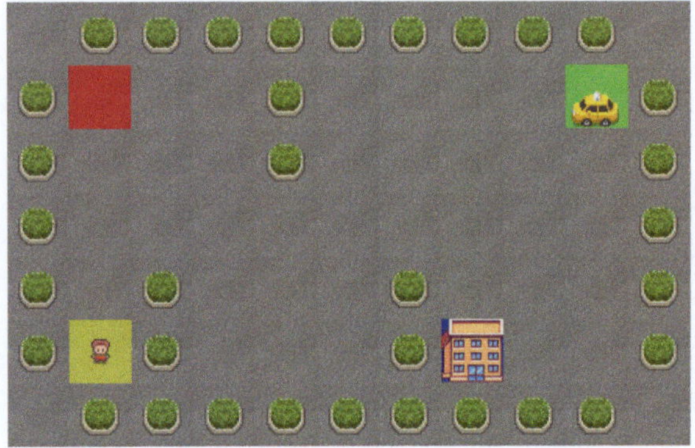

**Fig. 4.7** Rendering of the Taxi environment [61]

Nonetheless, given an instantiation of the environment, it is not possible for the action influence graph to predict the next operation that will be chosen by the trained agent. For this reason, it is necessary to couple it together with an approximation technique that is capable of predicting accurately the choices of the model-free RL algorithm. To accomplish this task, a variety of ML techniques can be applied to learn the structural equations of the state features and the action (linear regression, decision tree, multi-layer perceptron [149]) or directly the policy of the optimal agent (decision policy tree with bounded and unbounded depth [148]). The difference between the two directions of forecasting lies in their output: while in the latter scenario, only the action considered by the agent in a specific instantiation can be predicted, for the former case, the relationship between the state features and the actions can be modelled, given a clearer representation of the general MDP tackled. Moreover, by focusing only on the action that will be selected, the accuracy of the predictions can be improved significantly [148]. Hence, from this ML predictor, denoted as $\mathbb{D}$, it is possible to derive the action that will be executed by the RL agent, and the decision process can be reasoned through the action influence model.

Thus, it is now possible to declare the formal definition of the minimally complete explanations generated through an action influence model $\mathcal{M}$. In detail, the focus is first concentrated on the answers to "Why" questions.

**Definition 4.3**   ([149]) *A minimally complete explanation for "Why" question generated through a model $\mathcal{M}$ consists of a tuple* $(X_r = x_r,\ X_h = x_h,\ X_p = x_p)$, *where $X_r$ is the vector of reward nodes that is reached by the causal chain of the graph to sink nodes, $X_h$ is the vector of variables of the head node of the action $a$, predicted by the prediction model $\mathbb{D}$, $X_p$ is the vector of immediate predecessors of any variable in $X_r$, and $x_r, x_h, x_p$ give the value of the corresponding variables under the instantiation $\mathcal{M}_{\mathcal{X} \leftarrow x}$.*

Intuitively, given an action influence graph, it is possible to generate the explanation by looking at the chain of connections created by the specific action predicted by the chosen post-hoc ML algorithm. If multiple variables are encountered during this causal path, then only the source and destination nodes are considered, omitting all the intermediate vertices for writing the minimal explanation.

For example, considering the action influence graph in Fig. 4.6 and supposing that, in the same actual instantiation of Fig. 4.7 (South-East, South-West, North, East), the selected action is Go West. Then, from the graph, the head node $X_h$ = Destination can be identified as the vertex from which the edge labelled with the action Go West starts. The arrival node is instead denoted as the predecessor, meaning that $X_p$ = Taxi col.. In fact, following the subsequent path, it is possible to reach the reward variable $X_r$ = Passenger position. In this way, the causal chain can be recognized in the path generated by the vertices Destination → Taxi col. → Passenger position. Finally, the minimally complete explanation for answering to the question "Why Go West?" can be mathematically answered by showing the values of the actual instantiation for each variable in the causal chain: (Destination = South-East, Taxi col. = East, Passenger position = South-West). The complete explanation can be stated as follows:

"Because Destination = South-East, it is desirable to do action Go West to change Taxi col. = East as the goal is related to Passenger position = South-West".

This idea can be exploited also for answering the counterfactual queries, that is for the question "Why not Action B?". In this case, it is possible to compare the outcomes of the actual choice with the results when the action suggested is the counterfactual one. From this comparison, the differences, both in the causal chain and the predicted values, can highlight what the conditions, for which the actual action chosen by the agent is the best, are. Given the different context from the "Why" question, a proper definition of this minimally complete explanation has to be introduced. Below, the compact form of the minimally complete explanation is used applying the notation $(X = x)$ instead of the tuple $(X_r = x_r,\ X_h = x_h,\ X_p = x_p)$.

**Definition 4.4** ([149]) *Given a minimally complete explanation $(X = x)$ for an action a, predicted by the prediction model $\mathbb{D}$, under the actual instantiation $\mathcal{M}_{\mathcal{X} \leftarrow x}$, and a minimally complete explanation $(Y = y)$ for action b under the counterfactual instantiation $\mathcal{M}_{\mathcal{X} \leftarrow y}$, minimally complete contrastive explanation for "Why not" questions is defined as the tuple $(X' = x',\ Y' = y',\ X_r = x_r)$ such that $X'$ is the maximal set of variables in X where $(X' = x') \cap (Y' = y') \neq \emptyset$, when $x'$ is contrasted with $y'$; and $X_r$ and $x_r$ are as in Definition 4.3.*

Practically, the contrastive explanation proposed is based on the extraction of the differences between the causal chains created, following action *a* and *b*.

To clarify this definition, the usual example for the Taxi environment is used. Let the state and action instantiation be the same as the previous explanation, in this way the minimally complete explanation for the action Go West is already done. Then, the counterfactual action considered is Go South. Thereby, the minimally complete explanation for the counterfactual question can be created following the previously described approach, obtaining (Destination = South-East, Taxi row=North, Passenger position = South-West). Then, by comparing the two minimally complete explanations, it can be noticed how the shared variables are (Destination = South-East, Passenger position=South-West). Furthermore, by looking at the causal chain discovered in the action influence graph (Destination → Passenger position), the reward variable is identified in Passenger position. Therefore, the minimally complete contrastive explanation for "Why not Go South?" is formally answered by (Destination = South-East, Passenger position = South-West, Passenger position = South-West), which can be stated as:

"Because `Destination = South-East`, it is more desirable to do `Go West` instead of `Go South`, and change `Passenger position = South-West`, as the goal is related to `Passenger position = South-West`".

However, this particular counterfactual explanation can be improved if *opportunity chains* are introduced in the description. This improvement was proposed by the same authors in the paper [148]. In this work, Madumal et al. applied the concept of opportunity chains and *distal action* in the MDP scenarios. An opportunity chain can be described as an extension of the causal chain, presenting the following form: the variable $A$ enables $B$ which causes $C$. In this exemplification, the event $B$ is called a *distal event* or *distal action* and becomes the centre of the reasoning, due to its relevance for the completion of the task. For example, in the benchmark RL Taxi environment, where an agent has to learn to pick up a passenger and drive them to the destination, the obvious distal action is to pick up the customer. In fact, it has to be enabled by moving to the customer's location first, otherwise, the completion of the task is impossible. Nonetheless, also the action of dropping them off is distal since it must be passed through a precise sequence of states, i.e., picking up the passenger and then reaching the final destination. Therefore, the general definition can be stated as an action that has to be enabled by a specific sequence of actions that have to be performed. Thus, a distal action is an action that depends the most on the execution of the current action [148].

Due to this characteristic, the task of predicting the distal actions can be translated as a time series forecasting problem. In fact, the distal actions are dependent on the trajectory that has been performed by the agent until the current time step. Moreover, they can be estimated considering their time series version which can be derived as follows: for each time step of the trajectory analyzed, the next distal action can be associated as target value; then, when that particular distal action is performed, the target distal time series will update its current value with the subsequent distal action. For example, in the case of the Taxi environment, the target distal time series will be composed of `Pick up` actions for each time step until the passenger is inside the taxi car. Then, the target distal actions will be `Drop off` until the end of the episode. For this reason, the distal actions can be predicted through the adoption of Recurrent Neural Networks (RNNs), as suggested by [148]. This specific procedure will be resumed and explained extensively in the following sections.

Thereby, the previously introduced explanations for the counterfactual actions can be extended including also the distal action as a piece of information for the answer. By doing so, the *minimally complete distal explanation* is defined as follows:

**Definition 4.5** (Reformulated from [148]) *Given a minimally complete explanation $(X = x)$ for the current action a predicted by the prediction model $\mathbb{D}$, a minimally complete explanation $(Y = y)$ for the counterfactual action b and a distal prediction model $\mathbb{L}$, a minimally complete distal explanation for a "Why not" question is a tuple $(X_r = x_r,\ X_{con} = x_{con}, x_d)$, where $X_r$ does not change from Definition 4.3, $X_{con}$ is the maximal set of variables for which $X_{con} = (X_b = y_b) \cap (X_c = y_c)$, with $X_b$ being the intermediate nodes of the causal chain of the counterfactual action b, and $X_c$ are the variables that were modified by the particular choice of b, compared to the changes produced by a, and $x_d$ represents the distal action predicted through $\mathbb{L}$ s.t. $a_d \in \mathbb{A} \cap \mathbb{A}_c$, in which $\mathbb{A}_c$ indicates the set of actions in the causal chain created by the actual action a. All the values $x_r$, $x_{con}$ have to be contrasted using the actual instantiation $\mathcal{M}_{\mathcal{X} \leftarrow x}$ and the counterfactual instantiation $\mathcal{M}_{\mathcal{X} \leftarrow y}$.*

Due to the complexity of the mathematical formulation, it is helpful for comprehending the practical outcome of this approach to illustrate an example of an explanation. The circumstance analyzed is always the same: the actual action is Go West, while the counterfactual one is Go South. The corresponding minimally complete explanations are (Destination = South-East, Taxi col. = East, Passenger position = South-West) for the actual action and (Destination = South-East, Taxi row = North, Passenger position = South-West for the counterfactual one. Then, the variable between the head and reward variables of the counterfactual causal chain is Taxi row, so $X_b = $ Taxi row. Moreover, we can notice how this variable is also the one that has not been selected in the actual causal chain, whereas the intermediate variable Taxi col. was considered. Hence, it can be determined $X_c = $ Taxi row, leading to finding the contrastive variable $X_{con} = $ Taxi row. Moving to the selection of the distal action, as previously stated, it can be considered the action Pick up since the passenger is still waiting. Therefore, $a_d = $ Pick up. Finally, the reward variable considered by both the causal chains is $X_r = $ Passenger position. By completing the information with the actual and counterfactual instantiation, the minimally complete distal explanation for "Why not Go South?" question can be derived:

(Passenger position = South-East, Taxi row = North, Pick up).

The textual statement can be extended involving also the intrinsic information obtained by the optimal split of the decision policy tree and the correct intermediate variable of the actual causal chain. The final syntactical outcome is the following:

"Because `Taxi row` is at the optimal value `North`, it is more desirable to do `Go West` and change `Taxi col. = East`, instead of doing `Go South` and change `Taxi row = North`, to enable the action `Pick up`, as the goal is related to `Passenger position = South-East`".

The textual template used for the explanations above has been adapted to the Taxi context from the Starcraft environment presented in [148]. In the remaining part of this section, a critical analysis of the pitfalls in the methodology proposed by Madumal et al. [148, 149] is discussed.

## 4.2.2   Limitation of the Reviewed Methodology

The first and most clear issue in the methodology reported above involves the generation of the action influence graph. Those specific causal models must be hand-crafted only by an expert on the problem solved by an RL agent. Only in this way, it is possible to understand what the actions that are causally connecting some nodes are. Moreover, this choice can be biased by the human perception of the special situations encountered in the experience of the professional users. This aspect allows for the production of causal structures that depend on the mental representation of individual experts, leading to a non-well-defined and subjective solution. Additionally, it is required that an expert spends time on the development of this causal model, causing a possible waste of their precious hours that could be allocated differently and more productively. Besides the time component, the existence of experts in a particular problem can not always be guaranteed. From observations of non-expert users, erroneous causal representation can be deducted, invalidating all the well-defined mathematical constructs previously mentioned.

A correlated problem is the impossibility of automatically generating the explanations, given the requirement of a human supervisor for the preliminary step. Despite being completely automatic in the phase subsequent to the graph construction, there is still a consistent difficulty in developing algorithms that could generate an action influence graph directly from data. While multiple approaches deal with the topic of learning causal structures from observations, like SEMs and BNs, there are still no methodologies that can obtain similar results for the action influence models.

Along the same lines of automation, the distal actions that have been considered in [148] were also hand-crafted. Hence, this nullifies all the effort in the generation of a causal explanation. As it will be described in detail in Chap. 5, a wide set of importance metrics have been developed to detect the most useful states, in terms of both efficiency of the training and performances in increasing the explainability. From these data, the relevant actions are consequently deduced. Nevertheless, the problem of detecting distal actions is still open due to the impossibility of recognizing the major characteristics of these specific actions.

Besides the identification of the distal actions, another limiting facet is the usage of the distal concept for just the actions. Indeed, this quality can be straightforwardly extended to states and the final return. Thereby, the knowledge adopted in the creation of the explanations can be more accurate and detailed, leading to probable improvements in their comprehension. Furthermore, no counterfactual distal information has been contemplated. Although forecasting the counterfactuals in these situations can decrease even more the accuracy of the overall predictions, they can offer the users a direct comparison between the two actions chosen. Consequently, it can give insights into the different planning paths that could be taken.

From these limitations, this thesis introduces an approach that is completely automatic for the creation of a causal explanation and that includes the distal information regarding not only actions but also states and other fundamental values. The full methodology is illustrated in the next section.

## 4.3  BENEDICT: An Automatic Distal XRL Approach

As mentioned before, the major limitations, that arose from the analysis of the approaches stated in [148, 149], concern the lack of automation in the creation of the explanation and the limited amount of information regarding the distal aspects of a given problem. Therefore, a novel methodology that can generate causal explanations automatically and that includes a wider range of distal information is now proposed. The main idea lies in the combination of BNs with a distinct structure for the causal predictions and RNNs specialized for dealing with "Why" and "Why not" questions. For this reason, the following procedure has been named: Bayesian – rEcurrent neural NEtwork approach for Distal Information and Causal explanaTions (BENEDICT). An overview of the complete procedure suggested is shown in Fig. 4.8.

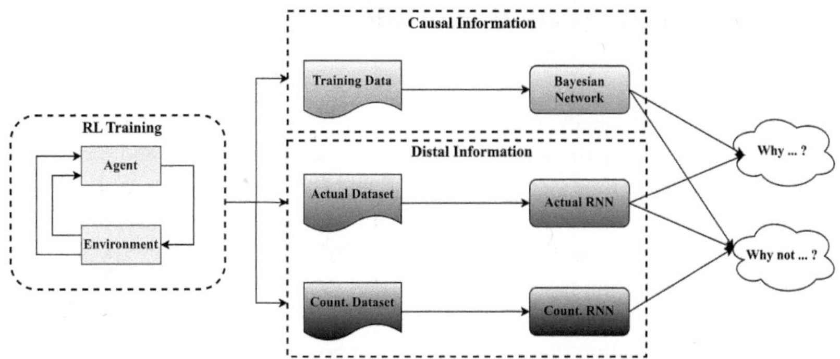


**Fig. 4.8** Pipeline of BENEDICT

The considered approach consists of four components. The first one is the trained model-free RL agent. In the computational experiments reported in Sect. 4.4, the details of the training phase and the particular choice of the selected RL method are discussed.

During the learning process of the agent, the trajectories are stored in a dataset that will be subsequently used in the creation of the causal structure. For the generation of causal explanations, a special structure of BN is considered. Besides the fact that these graphs can be learned efficiently from data, they can also guarantee a linear routine in the creation of the explanatory statement. Moreover, they can provide competitive levels of accuracy in forecasting [96]. The choice of adopting BN in the context of XRL is supported by these rationales.

For the distal elements in the explanation, two RNN units were created: one for addressing the predictions of the actual distal information and the other for the counterfactuals. For the training of these methodologies, a new dataset, composed of the trajectories saved after the training of the RL algorithm, was used. Combining the outputs of both the RNNs and the BN derived in the previous step, it is possible to create explanations for "Why" and "Why not" questions.

More specifics on processes leading to the production of the final components of the methodology are carefully reported in the successive sections.

### 4.3.1   Learning the Causal Structure

Fundamental for the creation of the DAGs of the BNs are the data saved during the training of the RL agent. The saved memory $M$ stores the actual state features $s_t$, the action chosen by the agent $a_t$, the next state features $s_{t+1}$, and the reward obtained $r_{t+1}$, at each time step $t = 0, \ldots, T$ and for each episode $e = 1, \ldots, Ep$ (here the dependency of the MDP values to the episode $e$ is clear, but in order to not overcrowd the notation, the $e$ subscript is avoided). This memory set is typical information stored during the training of various kind of RL agents, e.g., DQN, or modified versions of Q-learning [231]. The selection of these data is justified by the necessity of predicting the future changes in both the environment but also for forecasting the action chosen by the RL agent at the actual time step. To this end, the desired structure should present in-between characteristics of the classical BNs and their time-series versions, i.e., DBNs. Indeed, if each of the features in $M$ is associated with a node in the BN, then there are multiple time step nodes for the state components but also single time step vertices for the rewards and actions. By these means, if the structure is restricted to the state values, then it will look like a DBN, while in the general observation, it can be recognized as a BN. Considering those circumstances, the NOTEARS algorithm was selected as the first choice for the determination of the DAG.

Although the NOTEARS method can learn effectively the DAGs in universal contexts, in the scenario of MDPs it is possible to exploit some basic characteristics of the problems. In particular, some correction rules that can improve the quality of the constructed graph can be stated following the assumptions of being in an MDP:

- time direction: no future time step values can influence past characteristics;
- stationarity of dependencies: the relationships between state features are conserved at different time steps;
- importance of the actions: all the actual state features influence the choice of the action, and the action is determining also the future state features, implying that it will have incoming edges from actual time step state nodes and outgoing connections to the next time step state vertices.

These conditions can be implemented after the thresholding phase of NOTEARS, leading to an improved pruning process. The method with the above added characteristics will be denoted as the NOTEARS-MDP algorithm. In Algorithm 4.6 the full procedure is reported, where $s_{i,t}, \forall i = 1, \ldots, S_{feat}$ are the different features of the state vector $s_t$ with $S_{feat}$ indicating the number of state features, and the notation $E(B^*)$ denotes the edge set of the BN $B^*$

---

**Algorithm 4.6** NOTEARS-MDP algorithm

---

**Require:** Initial guess $(B_0, \alpha_0)$, observed data matrix $X \in \mathbb{R}^{m \times n}$, progress rate $c \in (0, 1)$, tolerance $\varepsilon_M > 0$, threshold $\varepsilon > 0$.

1: **for** $t = 0, 1, \cdots, \infty$ : **do**
2:    Solve the primal problem:

$$B_{t+1} \leftarrow \arg \min_{B \in \mathbb{R}^{n \times n}} (L^\rho(B, \alpha_t)) \text{ with } \rho \text{ s.t. } h(B_{t+1}) < c \cdot h(B_t).$$

3:    Dual ascent:

$$\alpha_{t+1} \leftarrow \alpha_t + \rho h(B_{t+1})$$

4:    **if** $h(B_{t+1}) < \varepsilon_B$ **then**
5:        Store the found solution

$$B^*_{approx} \leftarrow B_{t+1}$$

6:        **break**
7:    **end if**
8: **end for**
9: Thresholding:

$$B^* \leftarrow B^*_{approx} \circ 1_{[|B^*_{approx}| > \varepsilon]}$$

10: MDP-Pruning Conditions:
11: **for** $i = 1, \ldots, |S_{feat}|$ **do**
12:    Add if not already existing

$$(s_{i,t}, a_t) \in E(B^*)$$

                                                                 ▷ Actions importance

13:    **for** $j = 1, \ldots, |S_{feat}|$ **do**
14:        Add if not already existing

$$(a_t, s_{j,t+1}) \in E(B^*)$$

                                                                  ▷ Actions importance

15:        **if** $(s_{j,t+1}, s_{i,t}) \in E(B^*)$ **then**
16:            Cancel edge $(s_{j,t+1}, s_{i,t})$                             ▷ Time direction
17:        **end if**
18:        **if** $(i = 1) \wedge (s_{j,t+1}, a_t) \in E(B^*)$ **then**
19:            Cancel edge $(s_{j,t+1}, a_t)$                               ▷ Time direction
20:        **end if**
21:        **if** $(s_{i,t}, s_{j,t}) \in E(B^*)$ **then**
22:            Add if not already existing

$$(s_{i,t+1}, s_{j,t+1}) \in E(B^*)$$

                                                                    ▷ Stationarity

23:        **end if**
24:    **end for**
25: **end for**
26: **return** $B^*$

---

To demonstrate the quality of the proposed method, the outcomes of this methodology are evaluated for the Taxi environment through a comparison with the classical approaches reported in the previous section. Fig. 4.9 shows the DAG obtained from the modified algorithm NOTEARS-MDP with a threshold parameter $\varepsilon = 0.5$, while the results of the classical methodologies are reported in Fig. 4.10. Specifically, in order to highlight the differences between the classical NOTEARS results (Fig. 4.10a) and the NOTEARS-MDP (Fig. 4.9), the edges not shared between the two graphs have been coloured applying the following scheme: in green there are the edges that have been added and in red are the connections that have been eliminated using the novel pruning technique.

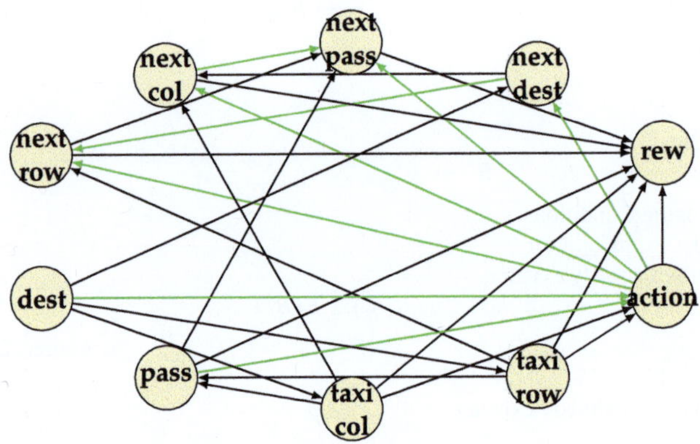

**Fig. 4.9** DAG obtained from NOTEARS-MDP with $\varepsilon = 0.5$

The first observation that can be drawn concerns the fundamental corrections that have been contributed by the pruning rules added. In fact, it is noticeable how the normal graph derived from NOTEARS (Fig. 4.10a) is individuating a causal connection from the next time step column and row position and the action variable. In addition to this fallacy, the influences of the taxi positions are not well recognized in the future time step features, leading to a missing link between the next column and the next passenger position nodes.

Due to the similarity in the temporal structure of the wanted BN, the results of NOTEARS-MDP are contrasted to the DBNs, obtained through DYNOTEARS using diverse threshold parameters, to identify possible inconsistencies. As can be seen in Fig. 4.10b and Fig. 4.10c, the outcomes of DYNOTEARS are satisfying only when a small threshold parameter is utilized $\varepsilon_A = \varepsilon_B = \varepsilon = 0.01$, implying that

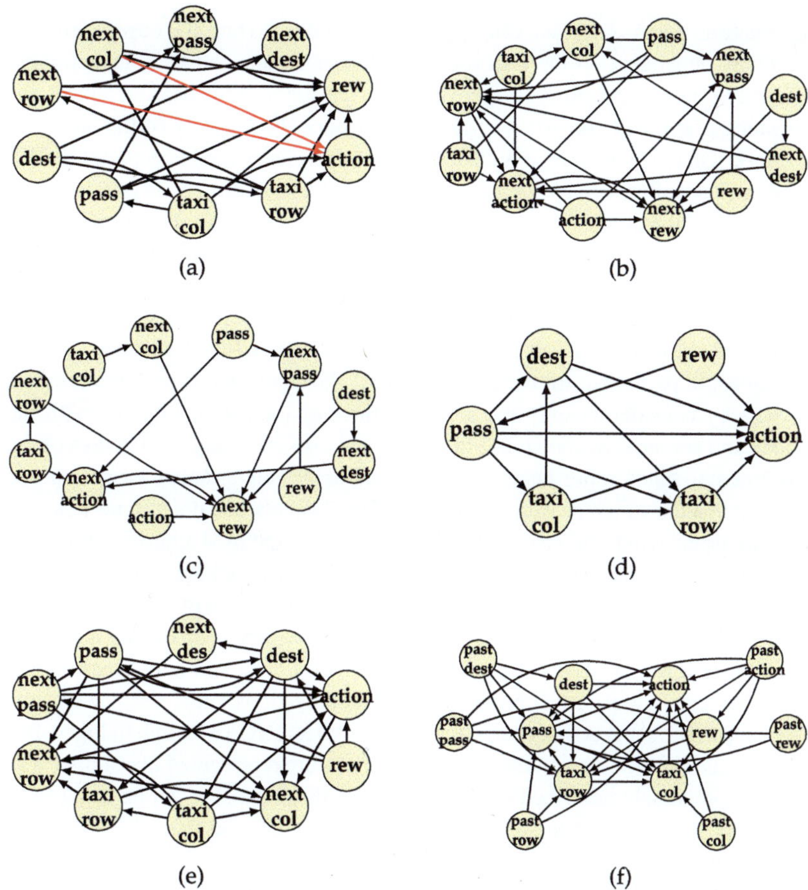

**Fig. 4.10** DAG obtained from classical algorithms: (a) NOTEARS with $\varepsilon = 0.5$, (b) DYNOTEARS with $\varepsilon = 0.01$, (c) DYNOTEARS with $\varepsilon = 0.1$, (d) DirectLiNGAM with actual features, (e) DirectLiNGAM with actual and future features, (f) VARLiNGAM with $\varepsilon = 0.1$

the structural equations of the causal graphs are shallow. Thus, the edges identified are highly dependent on the data used and diverse results can be obtained under different initial instantiations. Moreover, by increasing the threshold parameter, the remaining graph will consist of a sparse graph since the majority of the links will be erased. On the other side, it is clear how the edges discovered using NOTEARS-MDP are less influenced by the fluctuations of the data studied. This consideration is justified by the larger $\varepsilon = 0.5$ applied. Nevertheless, the causal connections that are found using DYNOTEARS are the same as the ones in NOTEARS-MDP, thus, endorsing the proposed methodology.

A comparison with the graphs created by the SEM-related algorithm such as DirectLiNGAM and VARLiNGAM (Figs. 4.10d–4.10e–4.10f), leads to interesting reflections. In particular, for the former method, the analysis was performed in two batches: by considering only the actual features of the states, together with the actions and rewards, as shown in Fig. 4.10d, and also including the features of the next time step states, represented in Fig. 4.10e. For the latter graph, it is necessary to point out that the variables were treated as different values, and not as time series. In fact, this aspect will be taken into account by the VARLiNGAM approach discussed successively. The main remark, which can be stated by looking at the DAGS of Fig. 4.10d and Fig. 4.10e, deals with the reversed edge present in the links between the rewards and the state features. This peculiar discovery is obviously a mistake since the rewards in MDPs depend on the previous state and action performed. Thus, they can at most affect possible causal modifications in the subsequent time step values. The same error is perpetuated in the VARLiNGAM graph in Fig. 4.10f. Despite this significant error, the other connections are represented in a similar way as in the previously described cases. Only a few discrepancies can be detected in the comparison between the output of NOTEARS-MDP and DirectLiNGAM for multiple time steps.

Along the same lines, the DAG created by VARLiNGAM, shown in Fig. 4.10f, shares the majority of the edges with the initial graph in Fig. 4.9. In detail, these are the only two graphs that can recognize the destination as a root node. From this vertex, the identified causal chain passes through the taxi column and row position, to finally end in the passenger position. The remaining issue, which can be outlined from the examination of the VARLiNGAM graph, deals with the difficulty in recognizing the variables connected to the reward. The only nodes having an outgoing edge pointing to the reward variables are the actual time step passenger position and action. Although these links are theoretically correct, they are not the only relevant information to estimate the reward. Also in this facet, the suggested graph outperforms the classical algorithm results since it can discern the complete set of features connected to the reward.

For all the reasons mentioned before, the innovative approach introduced here can be acknowledged as the best-performing one (for more results supporting this statement, see Appendix B.1 in the Electronic Supplementary Material). So, from now on, the causal structures will be constructed using the NOTEARS-MDP algorithm. After fixing the process for the generation of the DAGs, the focus is now moved to learning the probabilities according to the aforementioned graph structure and the training data.

### 4.3.2   Learning the BN Probabilities

To complete the inference of the BN probabilities, it is necessary to convert the information of the dataset into discrete values. In fact, while the procedure for learning the causal graph can be computed using discrete or continuous values, for learning the BN probabilities it is necessary to discretize the data. Only in this way, it is possible to exactly infer the probabilities from the observations stored. Nonetheless, this preliminary step can be avoided in already discrete environments, e.g., the Taxi problem.

In continuous scenarios, it is proposed a discretization exploiting the K-means clustering method (for more details see Appendix B.2 in the Electronic Supplementary Material). This algorithm was preferred to the usual tile encoding due to its ability to distinguish various dimensions clusters according to the distribution of the data points. Specifically, the tile encoding consists of dividing the continuous space into multiple overlapping tiles. Thereby, each point is associated to a sequence of 0 and 1 constructed following this rule: when a point is inside of tile number $i$, then it has at the $i$-th position in the sequence a 1; while, in the opposite case, if it is not in tile number $i$, there will be a 0 at the $i$-th location. Clearly, this method does not take into account the different distributions, thereby, the discretization applied is uniform. Instead, through the K-means clustering approach, combined with the analysis of the silhouette coefficients, it is possible to discriminate the regions where the majority of the points are, and then, split them into smaller portions to allocate a similar amount of data for each region. However, this division is strictly dependent on the number $K$ of clusters, so a detailed analysis has to be reported for each specific experiment.

Afterwards, having now discrete values, the Bayesian estimators' method is utilized to learn the probabilities of the BN. The principal assumption here is relative to the prior distribution being a Dirichlet distribution. As a consequence, the complete BN is formed. With this model, the future modifications in the environment and the actions decided by the agent can be quickly forecasted. The information generated in this manner will be central in the creation of the causal clause of the explanation. In the following, the formal definition of the actual and counterfactual predictions is declared.

**Definition 4.6** *Given an initial state $s_t \in \mathbb{S}$ at time step $t$, and an actual action $a_t \in \mathbb{A}$, the actual prediction $P_{act}$ using a BN $B = \{G_B, P_B\}$ is defined as follows:*

$$P_{act} = \arg \max_{s_{t+1} \in \mathbb{S}} P_B(s_{t+1}|s_t, a_t), \tag{4.66}$$

*which represents the state values maximizing the probability of realization, conditioned by the hypothesis of $s_t$ and $a_t$.*

*Similarly, given a counterfactual action $\bar{a}_t \in \mathbb{A}$ instead of the actual action $a_t$, then the counterfactual prediction $P_{count}$ using a BN $B = \{G_B, P_B\}$ is denoted as:*

$$P_{count} = \arg \max_{s_{t+1} \in \mathbb{S}} P_B(s_{t+1}|s_t, \bar{a}_t), \tag{4.67}$$

*where the conditioning is done by the counterfactual action $\bar{a}_t$ and the actual state $s_t$.*

For the actual prediction definition, it can be remarked the fact that also the actual action can be forecasted through the BN. Although this possibility is present and the accuracy of this task will be tested, it does not play a fundamental role in the creation of the explanatory sentence due to the already decided action to perform. In fact, at the moment of the question, it will be asked to reason for the choice of the actual action. Hence, this particular action can be directly used as input in the causal model created before.

In the next section, the attention can be redirected to learning the distal components of the full methodology and their exploitation for the generation of the distal clauses in the explanations.

### 4.3.3   Learning the Distal Information

The concept of distal actions was first introduced in [148], where it was defined as one of the most significant actions that have to be enabled through a precise sequence of actions. As already stated, this definition limits the application of this quality to only actions, while other values can benefit from this characterization. A straightforward extension is represented by the case of distal states, which can be recognized as states that have to be made available after a succession of visits of particular states. Scenarios like these are quite common in MDPs solved using RL agents. Consider, for example, the Taxi problem, where it is necessary to pass through a bottleneck in the centre of the map (highlighted in orange in Fig. 4.11) to move from one side to the other, or in the Key-Door problem (Appendix A.5 in the Electronic Supplementary Material), where, in order to reach the goal, it is required to pick up the key in a specific location.

**Fig. 4.11** Taxi environment with central bottleneck highlighted in orange

This characteristic can be shared as well with the final return of an episode, i.e., the accumulated rewards along a complete run of an environment. In fact, this value is strictly dependent on the exact sequence of states visited and actions selected: the optimal final return depends strongly on the execution of the correct actions for each time step, thus, it can be recognized the already described property of being *distal* reiterated for each decision time. Therefore, the final return can be included in the set of data that can be indicated as *distal information*. Moreover, this aspect is always provided in the same form for every environment, meaning that it can be used in the explanations no matter the problem addressed by the RL algorithms. In general, this is not the case for the distal actions or states, which are strictly related to special cases of MDP. However, the scenarios where the distal actions and states do not exist can be restricted to particularly simplified environments, e.g., when there are less than or equal to two variables. Nonetheless, information about the final return can additionally improve the quality of the insights reported in the explanations by estimating the overall outcome. Through a comparison of the multiple counterfactual situations, the optimal choice can be immediately recognized.

Thus, the concept of distal counterfactual information that can be contrasted to the distal actual information can be derived. However, the whole reasoning is based on the hypothesis that these data can be predicted accurately.

To accomplish this task, two RNNs were trained in predicting the distal actual information and the distal counterfactual information. For the learning phase, an extended memory dataset was created by storing the trajectories of the fully-trained RL algorithms. The new memory saved consisted, at each time step $t$, of the state features $s_t$, actions performed $a_t$ and the Q-values associated with the state-actions couples visited $q_t = Q(s_t, a_t)$. When the Q-values were not available since a policy gradient method was applied, they were replaced by the accumulated rewards until that time step. In this way, the overall process can be generalized to any model-free approach. Nevertheless, due to the splitting for the actual and counterfactual forecasting, the datasets were also characterized by two different ways of selecting the values to save. While the actual dataset is produced by following the optimal policy and keeping all the values at each time step, the counterfactual one was obtained through the simulation of a counterfactual choice. Specifically, the agent samples actions optimally until a particular random time step is reached. In that instant, the selected operation will be different from the choice of the trained RL algorithm. Subsequently, all the remaining actions will be always following the optimal solution. Through this strategy, it is possible to estimate the consequences of the wrong action, hence, reporting the proper modifications caused.

After identifying the proper set of data that has to be used, it is required to pre-process them to facilitate the learning of the RNNs. Particularly, two operations, next to the feature extractions procedure, are executed: first, at each time step, the actions are one-hot-encoded and the Q-values are normalized to decrease the time of the training phase and provide comparable values for the losses utilized.

The second process that has to be applied is padding the time series with zeros, in order to have the same length vectors. In fact, at the start of each episode, the saved values $v$ are only composed of the initial data, so $v \in \mathbb{R}^{1 \times f}$, where $f$ represents the total number of features stored. At the end of the episode, instead, the dimensions of the stored memory have length $l$, implying that the vector $v$ is saved as $v \in \mathbb{R}^{l \times f}$. Hereby, each data point of the obtained datasets are time series with different lengths. Despite this incongruity, it is possible to rectify the dimensions for each of them by adding the correct number of vectors of zeros. To this aim, the maximum length $M$ time series has to be detected. Thereby, the data points will be filled with zero vectors at their beginning, so no issues can be provoked. The main phases for the generation of these two datasets are summarized in Fig. 4.12.

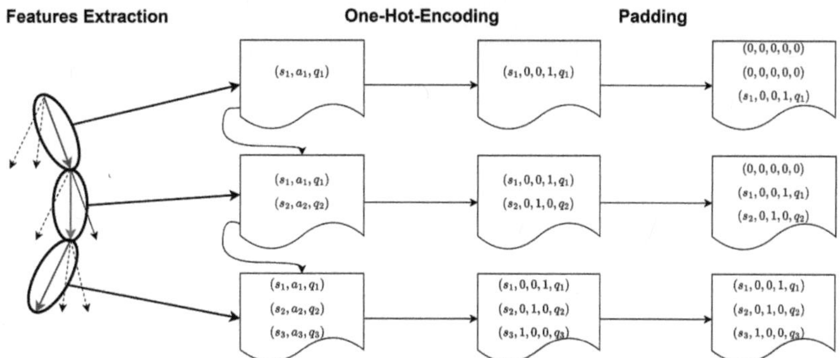


**Fig. 4.12** Visualization of the different phases of the dataset generation process

From these data, the actual RNN and counterfactual RNN can be trained to predict the next distal action or state and the final return, as suggested by Madumal et al. [148]. For the forecasting of distal information, a preliminary test on the architecture to adopt was carried out. The performances of the long short-term memory cells (LSTM) and gated recurrent units (GRU) are compared to determine the best option (for more details on these types of RNNs, see Appendix B.3 in the Electronic Supplementary Material). Four different combinations of RNNs were evaluated in the task of predicting the distal actions and the final return in the Taxi environment. The features used were the state features and actions selected, and on one side the Q-values and on the other the accumulated rewards until that time step. For each of these conditions, an LSTM and a GRU network, composed of a single layer comprising 50 cells, were trained over 100 epochs. The specific choice of the single architecture with multiple nodes and the numbers of epochs can be justified by the computational experiments already performed in this exact task [148] and similar ones [91]. The loss considered for the distal action predictions was the sparse categorical cross-entropy, while for the return the mean squared error was used. The training and test data (with a ratio 80 : 20), used for the training of the networks above, have been obtained from the completion of 1000 episodes for the Taxi environment. For more details about the parameters chosen for the RNNs, refer to Appendix B.4 in the Electronic Supplementary Material.

The results of this learning process are shown in Fig. 4.13. From Fig. 4.13a, can be recognized how all the combinations can perform the estimation of the final return efficiently. Nonetheless, a major difference is present between the methods with the Q-values as input and the ones with only the actual return. Despite the fact that the choice of Q-values can provide better approximations than the accumulated rewards, the availability of the Q-value function is not always verified. So, its application can be substituted with the actual return with only minor losses in the predictions. Moreover, no differences are spotted between the LSTM and GRU networks in both Fig. 4.13a and Fig. 4.13b. Thus, the latter architectures are recognized as the best options since they require fewer parameters than LSTM.

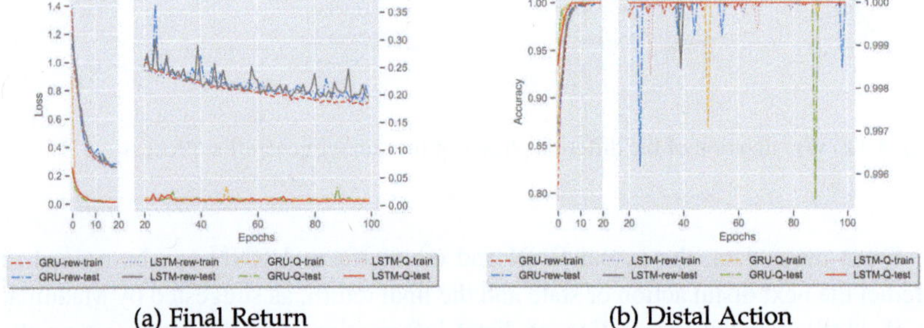


**Fig. 4.13** Comparison of losses and accuracy of RNNs with different cells and features for the Taxi problem

As previously done for the causal component, the formal definition for the distal actual and counterfactual predictions using the aforementioned GRU networks are declared below:

**Definition 4.7** *Given a GRU network* $f_{GRU}$ *and a set of padded vector of the states* $s_t$, *one-hot-encoded actions* $a_t$, *and corresponding Q-values* $q_t$ *(or accumulated rewards* $g_t$*)* $\forall t = 0, \ldots, M-1$, *where M is the maximum length of the time series, the distal actual prediction* $d_{act}$ *is defined as:*

$$d_{act} = f_{GRU}(\{s_t, a_t, q_t\}_{t=0}^{M-1}), \tag{4.68}$$

*with* $\{s_t, a_t, q_t\}_{t=0}^{M-1}$ *denoting the time series of the elements from* $t = 0$ *until* $t = M - 1$. *Similarly, given the counterfactual action* $\bar{a}_{M-1}$, *it is defined the distal counterfactual prediction* $d_{count}$ *as follows:*

$$d_{count} = f_{GRU}(\{s_t, \bar{a}_t, q_t\}_{t=0}^{M-1}), \tag{4.69}$$

*where* $\bar{a}_t = a_t, \ \forall t < M - 1$.

It is fundamental to observe that in Definition 4.7 the already padded vectors are used, implying that some of the values at the beginning time steps are equal to 0. Through this decision, it is possible to equivalently identify all the moments of asking the questions as the last time step $(M - 1)$ in the time series.

Definition 4.7, together with Definition 4.6, will be used in the pipeline for the creation of the explanations listed in the following section.

### 4.3.4   Creating the Explanations

So far, it has been described how to train and use the models for the causal and distal predictions. In this last methodological section, the procedure for delineating the explanations for both "Why" and "Why not" questions is discussed.

The first concept that has to be introduced is relative to the BN: the concept of *central variables*. This particular node will assume the role of the central vertex of a causal chain in the DAG of the BN. From this feature, all the explanatory statements will be consequently developed following a similar process to the one described in [148, 149]. Formally, the actual central variable is defined below:

**Definition 4.8**   *Given the state $s_t$ and the actual prediction $P_{act}$ performed by a BN $B = \{G_B, P_B\}$, an actual central variable (or node) $D \in V(G_B)$ is a node in the set of vertices $V(G_B)$ of the DAG $G_B$ such that:*

$$s_{D,t} \neq P_{D,act}, \tag{4.70}$$

*where $s_{D,t}$ and $P_{D,act}$ are respectively the D components of the state $s_t$ and predicted vector $P_{act}$.*

*Similarly, given the counterfactual prediction $P_{count}$ instead of the actual one, a counterfactual central variable (or node) $C \in V(G_B)$ can be defined as a node in the set of vertices $V(G_B)$ of the DAG $G_B$ such that:*

$$s_{C,t} \neq P_{C,count}, \tag{4.71}$$

*where $s_{C,t}$ and $P_{C,count}$ are respectively the C components of the state $s_t$ and predicted vector $P_{count}$.*

The definition above is relative to a singular feature that is changing between the actual state instantiation and the predicted future values. Nevertheless, in particular situations, multiple central variables can be spotted. For example, when the predicted values are all different from the actual state, then all the states' features will be identified as central variables.

After characterizing the central nodes for the causal chain as the features that will be modified in the next step, according to the predictions of the BN, it is possible to discuss the practical scopes of this piece of information. To help understand the selection and usage of the actual and counterfactual central variables, a practical example always relative to the Taxi environment is reported in Fig. 4.14. The situation analyzed is the same as the one shown in Section 4.2.1. In this case, the actual central variable is given by $D =$Taxi column since through the action Go West the BN correctly predicted a change in Taxi column position from East to East-Center. In the counterfactual scenario, the modification forecasted is relative to the taxi row that may pass from North to North-Center. Hence, the counterfactual central variable is the $C =$Taxi row.

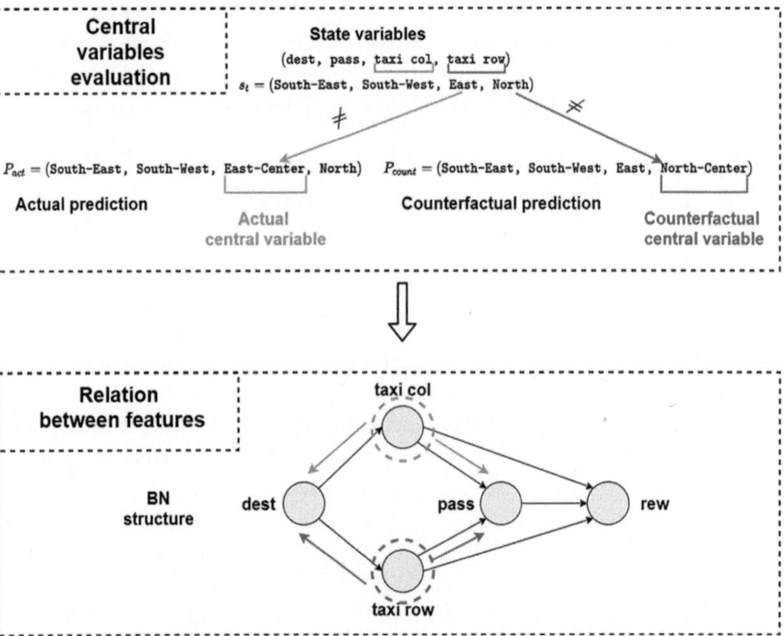



**Fig. 4.14** Procedure for the recognition of the actual and counterfactual central variables and their application in the identification of the causal chains

After detecting the central variables, it is fundamental to study the structural relationships of these features. Specifically, the nodes that will be used in the explanations are the parent and children vertices of the central variables in the subgraph including only the actual state features and the reward node. In the recurrent example, the parent nodes of the central variables are $Pa(D)$ =Destination and the child node is $Ch(D)$ =Passenger position, as visualized in the bottom part of Fig. 4.14.

The last characterization, that is needed before explicitly reporting the mathematical definition of the explanations, involves the reward features. In Madumal et al. [148, 149], the sink nodes were defined as the ones that were strictly related to the earnings and the completion of the goal. On the other hand, through the adoption of the reward as node in the BN structure, it is already possible to determine what are the variables inducing major changes in the rewards. In fact, this aspect can be discriminated by identifying the variables that are connected with an edge to the reward vertex. Then, the characteristics, that are part of the causal chain and that have an outgoing link to the reward vertex, are declared as the *reward features (nodes) $R(D)$* for the proposed BENEDICT methodology. In the mathematical formulation, the dependence of the reward features to the central variable is explicitly referred since it generates the causal chain from which the nodes connected to the reward vertex are selected.

Following the example above, the only variable that is in the causal chain and that has an edge connecting it to the reward vertex is Passenger position. Thus, Passenger position is declared as a reward node, i.e., $R(D)$ =Passenger position.

From all these data, the final explanations to the question "Why action $a_t$?" and "Why not action $\bar{a}_t$?" can be outlined by using the different information of the instantiations and predictions. Although the main ideas for generating the explanations have been reported in the previous example, it is significant to report the formal definitions to clear any doubts.

**Definition 4.9**  *Given the BN $B = \{G_B, P_B\}$, a GRU network $f_{GRU}$, a state $s_t$, an actual action $a_t$, the explanation for the "Why action $a_t$?" question is a tuple:*

$$(s_t, \ a_t, \ P_{act}, \ d_{act}, \ Pa(D), \ Ch(D), \ R(D)), \tag{4.72}$$

*where $P_{act}$ is the predicted future values for the state features obtained from the BN $B$, $d_{act}$ is the distal actual information forecasted by the GRU network $f_{GRU}$, $Pa(D)$ and $Ch(D)$ are respectively the parent set and children set of the central* *actual variable D, and R(D) is the set of reward nodes along the causal chain derived from D.*

The textual template used for practically translating the mathematical definition into a readable statement is the following:

> "Since *parent nodes of the actual central variable ($Pa(D)$)* is *instantiation of the parent nodes ($s_{Pa(D),t}$)*, it is desirable to do action *actual action ($a_t$)* in order to change *actual central variable ($D$)* from *instantiation of the actual central variable ($s_{D,t}$)* to *prediction of the BN ($P_{D,act}$)* to influence the *children nodes of the actual central variable ($Ch(D)$)*, because *rewards nodes ($R(D)$)* are connected to the goal. Following this action, it will be possible to do *distal action* and obtain a final return of *actual prediction of the return ($d_{act}$)*)".

Thus, the semantic explanations are generated by starting the reasoning from the actual instantiation of the parent set nodes. After the expression of these values, it is possible to recognize their influence in selecting the actual action to change the central actual variable. In this way, not only the central variable will be modified in the next time step (according to the predictions of the BN) but also the nodes that are in the children set will be affected. This particular choice will also be motivated by the variables that are considered as the reward nodes and by the predictions of the distal information.

In the case of the "Why not" question, the structure is slightly modified to include also the different outcomes related to the counterfactuals. Before explaining the details of these sentences, the following definition is necessary.

**Definition 4.10**  *Given the BN $B = \{G_B, P_B\}$, an actual GRU network $f_{GRU}$, a counterfactual GRU network $\bar{f}_{GRU}$, a state $s_t$, an actual action $a_t$, a counterfactual action $\bar{a}_t$, the explanation for the "Why not action $\bar{a}_t$?" question is a tuple:*

$$(s_t,\ a_t,\ \bar{a}_t,\ P_{act},\ P_{count},\ d_{act},\ d_{count},\ C,\ Pa(D),\ Ch(D),\ R(D)),  \qquad (4.73)$$

*where $P_{count}$ is the predicted future values for the state features obtained from the BN $B$ using the counterfactual actions $\bar{a}_t$, $d_{count}$ is the distal counterfactual information forecasted by the GRU network $\bar{f}_{GRU}$, $C$ is the counterfactual central variable and the remaining elements are as in Definition 4.9.*

The template considered for the counterfactual explanations defined above is the following:

"Since *parent nodes of the actual central variable* $(Pa(D))$ is *instantiation of the parent nodes* $(s_{Pa(D),t})$, it is more desirable to do action *actual action* $(a_t)$ in order to change *actual central variable* $(D)$ from *instantiation of the actual central variable* $(s_{D,t})$ to *actual prediction of the BN* $(P_{D,act})$ instead of changing *counterfactual central variable* $(C)$ from *instantiation of the counterfactual central variable* $(s_{C,t})$ to *counterfactual prediction of the BN* $(P_{C,count})$ doing action *counterfactual action* $(\bar{a}_t)$, to influence *children nodes of the actual central variable* $(Ch(D))$, because *rewards nodes* $(R(D))$ are connected to the goal. Following this action, it will be possible to do *distal actual action* and obtain a final return of *actual prediction of the return* $(d_{act})$, while following the counterfactual action will lead to *distal counterfactual action* and obtain a final return of *counterfactual prediction of the return* $(d_{count})$".

The main difference that characterizes the "Why not" answers consists in the comparisons derived from the counterfactual predictions. Indeed, it is possible to compare not only the corresponding central variable recognized but also the distal information predicted. Through this contrastive construction, the choices of the actual actions are reported following the same path as in the "Why" answers.

In this section, the focus was on explaining the methodologies that are behind the creation of reasoning clauses for actual and counterfactual scenarios. In the next part of the chapter, attention is moved to the computational experiments and the practical results that have been achieved by applying the BENEDICT approach. The evaluation tests were performed not only at the application level but also in the human perspective. Thus, in the next section, the efficiency of this proposed procedure has been validated in multiple environments while, in the subsequent part, the quality of the explanations generated will be proved.

## 4.4  Computational Experiments

So far, the results of the BENEDICT method have been stated only in specific cases of the Taxi environment. Despite this being one of the main benchmarks used in the literature for evaluating RL methodologies, a large variety of problems can be addressed. The proposed approach is tested not only in the discrete environment of Taxi but also in continuous situations such as Mountain Car and Cartpole. For each of these problems, a specific RL algorithm was trained until the task was fully learned. The particular choice of the combination of RL agents to adopt for each environment was decided by taking into account the analysis performed in the same settings as Madumal et al. [148, 149]. Thus, the performances achieved by BENEDICT and the other methodologies already applied in the literature can be lastly compared. In the following, the results obtained in each problem are delineated in depth.

### 4.4.1   Taxi Environment

The first environment discussed is the one that has been used before for the explanations of the methodologies of Madumal et al. [148, 149], and for clarifying some aspects of BENEDICT: the Taxi. This problem has been already introduced in Section 4.2.1, where an example of rendering was shown in Fig. 4.7. To solve this problem, the SARSA method was used. For more information of the training and the parameters used in it, please refer to Appendix B.4 in the Electronic Supplementary Material.

The dataset used for the creation of the BN is composed of 5000 episodes (runs) during the training phase. The DAG generated using the NOTEARS-MDP algorithm with threshold parameter $\varepsilon = 0.5$ was reported before in Fig. 4.9. After learning the probabilities associated with the graph and the data, the quality of the predictions of the actions and future state features is analyzed. For the study of the accuracy, 100 episodes were completed with the fully trained agent and the different actions and modifications of the environment were compared to the forecasted values from the BN. Moreover, also the AUROC score [36] is calculated for providing an alternative estimation of the general efficiency of the BN. The results, listed in Table 4.1, confirm the optimal predictions generated through the BN, achieving a nearly perfect recognition of the actions and changes in the state features.

**Table 4.1**  AUROC and accuracy for the prediction of each future feature in the Taxi environment

| Parameter | Action | Column | Row | Passenger | Destination |
|---|---|---|---|---|---|
| AUROC | 0.857 | 0.996 | 1.000 | 1.000 | 1.000 |
| Accuracy | 0.994 | 1.000 | 1.000 | 1.000 | 1.000 |

In the distal part, the only information considered were the actions of picking up and dropping off the passenger, and the usual final return earned. For the selection of the correct distal actions at the proper time steps, a rule-of-thumb was applied: before arriving at the client, "pick up" is associated as a distal action, while all the subsequent time steps will be connected to the distal action "drop off". Also in this case, the parameters used for the training of the actual and counterfactual GRU networks are reported in Appendix B.4 in the Electronic Supplementary Material. The training and test losses and accuracy computed during both phases are displayed in Fig. 4.15. In detail, Fig. 4.15a shows the loss for the final return and the accuracy for the distal action prediction. Whereas, in Fig. 4.15b, the outcomes for the counterfactual scenario are depicted. In both illustrations, it is possible to notice how the RNNs are able to accomplish the task of estimating the final return and whether the next fundamental action will be picking up or dropping off the customer. Furthermore, it is valuable to tell that no hyperparameter optimization was performed to improve the quality of the training. By this means, the correct prediction of the distal information can be achieved without any large search in the parameter spaces.

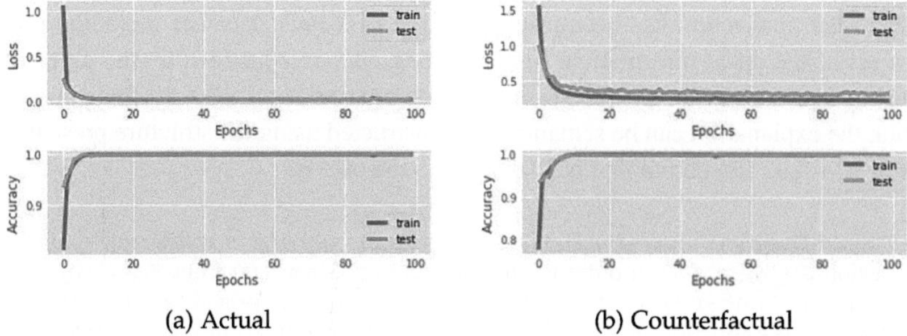

(a) Actual     (b) Counterfactual

**Fig. 4.15** Training and test losses and accuracy for the Actual and Counterfactual RNNs in the Taxi environment

After discussing of the overall efficiency for the predictions of the causal and distal information, it is now helpful to present a practical example of explanations for both the "Why" and "Why not" questions.

Beginning with the question about the actual action chosen, let the actual instantiation for the state features be $s_t$ = (South-East , South-West, East, North), where the first component denotes the destination, then, in order, the passenger and finally the column and row of the taxi position. The action selected by the SARSA agent is $a_t$ = Go West. The first step consists of predicting the next time step state features through the BN. The actual predictions produced are as follows:

$$P_{act} = \text{(South-East, South-West, East-Center, North)}. \tag{4.74}$$

From these values, it is straightforward to detect the central variable by comparing the prediction vector with the actual state: the taxi column is the only feature that is changing between the initial instantiation $s_t$ and the prediction $P_{act}$. By looking at the DAG of the BN, it is possible to identify the variables that are in the parent, children and reward set. Explicitly, the destination is a parent node, the passenger position is a child node and the reward nodes are both the passenger location and taxi column. Then, it is fundamental to calculate the estimations on the final return and the distal action linked to this position and action with the actual GRU network. The output generated is the following:

$$d_{act} = (\texttt{Pick up}, 4),\tag{4.75}$$

where the final return has been rounded to the next integer before including it in the set of the distal information. This rounding choice is justified by the usage of only integer values for the reward in the Taxi environment. With all the gathered data, the explanation can be semantically constructed using the structure presented in Sect. 4.3.4. The output that is visualized to the user is:

"Since $\texttt{Destination}$ $(Pa(D))$ is $\texttt{South-East}$ $(s_{Pa(D),t})$, it is desirable to do action $\texttt{Go West}$ $(a_t)$ in order to change $\texttt{Taxi column}$ $(D)$ from $\texttt{East}$ $(s_{D,t})$ to $\texttt{East-Center}$ $(P_{D,act})$ to influence the $\texttt{Passenger position}$ $(Ch(D))$, because $\texttt{Passenger position}$ and $\texttt{Taxi column}$ $(R(D))$ are connected to the goal. Following this action, it will be possible to do $\texttt{Pick up}$ and obtain a final return of 4 $(d_{act}))$".

The procedure is similar also in the case of the question "Why not $\texttt{Go South}$?". In this case, the additional tasks concern the counterfactual predictions involving the counterfactual action $\bar{a}_t = \texttt{Go South}$. Therefore, given the previous information, it is now relevant to include the predictions for the causal and distal parts using the corresponding counterfactual models. For the next time step state features, the predictions are always made through the same BN with the only exception that the new input will be $\bar{a}_t$. The output produced:

$$P_{count} = (\texttt{South-East, South-West, East, North-Center}),\tag{4.76}$$

where, in this situation, the central counterfactual variable is represented by the taxi row, since it is the only feature modified from the instantiation $s_t$. The remaining data that has to be computed is the distal counterfactual information through the counterfactual GRU network. The result of this operation leads to:

$$d_{act} = (\texttt{Pick up}, 1),\tag{4.77}$$

where it can be noted how the final return recognized is slightly different from the one that could be gained by following the optimal policy after the counterfactual decision. In this latter case, the final return would have been 2 instead of the approximated value of 1. Despite not being completely exact, this estimation can be still considered satisfying given the rounding step that has been applied before storing this data. In fact, the output of the RNN can be closer to the real value for some decimals ($< 0.5$). All the pieces of information extracted from the BENEDICT methodology are now compacted in a proper explanation for the "Why not Go South?" question. The resulting statement is shown below:

> "Since Destination $(Pa(D))$ is South-East $(s_{Pa(D),t})$, it is more desirable to do action Go West $(a_t)$ in order to change Taxi column $(D)$ from East $(s_{D,t})$ to East-Center $(P_{D,act})$ instead of changing Taxi row $(C)$ from North $(s_{C,t})$ to North-Center $(P_{C,count})$ doing action Go South $(\bar{a}_t)$, to influence Passenger position $(Ch(D))$, because Passenger position and Taxi column $(R(D))$ are connected to the goal. Following this action, it will be possible to do Pick up and obtain a final return of 4 $(d_{act})$, while following the counterfactual action will lead to Pick up and obtain a final return of 1 $(d_{count})$".

The testing environment examined here was the only one where no discretization had to be made due to the original discrete formulation of the Taxi problem. In the remaining two experiments, it will be discussed how to deal with the continuous state spaces.

### 4.4.2   Cartpole Environment

The first continuous environment investigated is Cartpole. For a detailed description of this environment, it is referred to Appendix A.2 in the Electronic Supplementary Material. To solve this problem, the Policy Gradient method REINFORCE was applied. Information regarding the parameters and the training data are reported in Appendix B.4 in the Electronic Supplementary Material.

Before creating the BN from the data collected during the learning phase, the dataset has to be discretized due to the continuous state features. To this end, the K-Means clustering technique was applied and, through the elbow method, the optimal number of clusters for each state component was found. For the sake of conciseness, the details relative to this analytical part are discussed in Appendix B.5 in the Electronic Supplementary Material. After this preprocessing process, it is possible to employ the NOTEARS-MDP as in the Taxi environment to detect the DAG associated with the stored training trajectories. The resulting graph is shown in Fig. 4.16.

The BN produced is evaluated in a prediction test consisting of forecasting the future state values and the decision made by the REINFORCE algorithm for solving the Cartpole environment in 100 episodes. As done previously in the Taxi problem, the metrics checked are the AUROC and accuracy in discriminating the correct action or discrete bin in which the state components will be. The results of this study are listed in Table 4.2. In this case, due to the discretization of the environment characteristics, the performances in the predictions are decreased. In particular, it becomes harder to recognize in a precise manner the velocity of both the pole and the cart since they can reach disparate peaks. Moreover, small changes in the trajectories followed could provide great modifications in the velocities, meaning that these values could be particularly complicated to anticipate. Despite reporting low accuracy in velocity forecasting, the AUROC values associated with these characteristics are instead proving a general optimal solution. Indeed, other than the value of the action, the AUROC of all the remaining features achieve satisfying levels (> 0.90). This discrepancy can also indicate that the stored episodes are not well generalizing the overall situations appearing in the Cartpole environment. By these means, the occasions where the BN performs worse are more present in the memory saved.

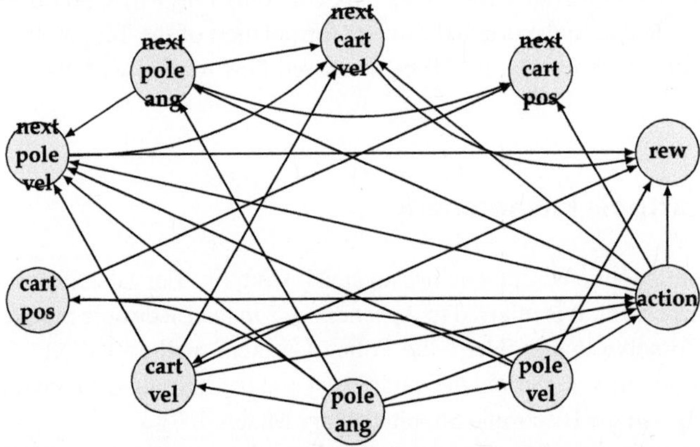

**Fig. 4.16** Structure of the BN for the Cartpole environment. The names of the nodes represent the following features: "cart pos" is the cart position, "cart vel" is the cart velocity, "pole ang" is the pole angle, "pole vel" is the pole velocity, "action" is the action, "rew" is the reward, "next cart pos" is the next value for the cart position, "next cart vel" is the next value for the cart velocity, "next pole ang" is the next pole angle, "next pole vel" is the next value for the pole velocity

**Table 4.2** AUROC and accuracy for the prediction of each future feature in the Cartpole environment

| Parameter | Action | Cart position | Cart velocity | Pole angle | Pole velocity |
|---|---|---|---|---|---|
| AUROC | 0.804 | 0.997 | 0.927 | 0.983 | 0.946 |
| Accuracy | 0.767 | 0.988 | 0.582 | 0.917 | 0.652 |

For the second phase of this analysis, the training of the GRU networks for the distal actual and counterfactual questions is described. As aforementioned in the discussion about the existence of the distal actions or states in any environment, in the Cartpole problem, no clear distal information other than the final return can be detected directly. The simplicity of the task indeed does not help in the definition of particular occurrences that should be enabled by specific chains of actions or states visited. For this reason, only the final return was outputted from the RNNs. Another singularity of this test is the lack of the state-action value function. Nevertheless, it has already been shown in the methodological part that the accumulated reward until the time step of the evaluation can be equivalently used as a replacement for the Q-value. The dataset for the training of the GRU networks is created by exploiting the trained REINFORCE agent and saving the first 100 time steps in each trajectory. This truncation was necessary for limiting the dimensions of the time series adopted. Alternatively, the memory required for the learning process is impracticable, leading to an unacceptable slowdown in the overall procedure. Even though the information is bounded in the time component, it can still summarize the fundamental data and efficiently train the GRU networks in both the actual and counterfactual scenarios, as shown in Fig. 4.17.

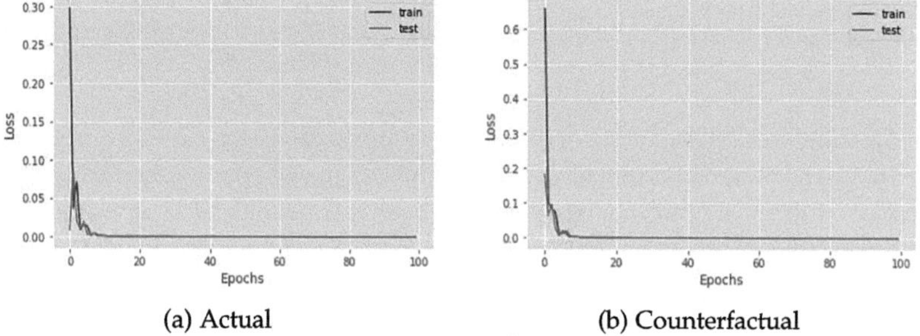

(a) Actual        (b) Counterfactual

**Fig. 4.17** Training and test losses for the Actual and Counterfactual RNNs in Cartpole environment

Finally, the full predictions are used for defining the explaining statement for the "Why" and "Why not" questions. An example for each of these situations is now discussed. Consider as actual state $s_t$ = (Center-Left, Low-Negative, Center-Right, Low Positive), where the first component denotes the cart position, then, in order, the cart velocity, the pole angle, and, lastly, the pole velocity. The REINFORCE agent decided to act $a_t$ = Go Left so the query to address is "Why Go Left?". Following the usual BENEDICT pipeline, as an initial step, the predictions in the next time step of the state features are calculated through the BN:

$$P_{act} = \text{(Center-Left, Low-Negative, Center-Right, Positive)}. \tag{4.78}$$

The actual central variable recognized in this case is the pole velocity. The remaining nodes from the causal chain derived from the pole velocity vertex are the pole angle as a parent node and the cart velocity as a child node. Between all of these features, the pole and cart velocities are the ones that have been linked to the reward node in the BN shown in Fig. 4.16. Then, the actual GRU network is employed to obtain the distal information, i.e., the final return:

$$d_{act} = 100. \tag{4.79}$$

The set of information needed for the creation of the answer to the actual question is now complete. Thus, the output of the BENEDICT methodology is the following:

"Since Pole Angle ($Pa(D)$) is Center-Right ($s_{Pa(D),t}$), it is desirable to do action Go Left ($a_t$) in order to change Pole Velocity ($D$) from Low Positive ($s_{D,t}$) to Positive ($P_{D,act}$) to influence Cart Velocity ($Ch(D)$), because Pole Velocity and Cart Velocity ($R(D)$) are connected to the goal. Following this action, it will be possible to obtain a final return of 100 ($d_{act}$)".

On the other hand, when considering the counterfactual action $\bar{a}_t$ = Go Right, the additional predictions that have to be completed include the counterfactual predictions of the changes in the state and the counterfactual distal information. For the former, the BN with the initial state instantiation and the counterfactual action as input is used to determine the future values of the state features. The evaluation of the computed probabilities led to the following outcome:

$$P_{count} = \text{(Center-Left, Low-Negative,}$$
$$\text{Center-Right, Low Negative),} \qquad (4.80)$$

where the modification still appeared in the pole velocity. Therefore, the actual and counterfactual central variables are the same. For the distal computation, the counterfactual GRU network is utilized in the estimation of the final return achieved, indicating the distal value:

$$d_{count} = 100. \qquad (4.81)$$

From the gathered data, the produced answer to the question "Why not Go Right?" is outlined by applying the same schematic semantic structure:

"Since Pole Angle $(Pa(D))$ is Center-Right $(s_{Pa(D),t})$, it is more desirable to do action Go Left $(a_a)$ in order to change the Pole Velocity $(D)$ from Low Positive $(s_{D,t})$ to Positive $(P_{D,act})$ instead of changing the Pole Velocity $(C)$ from Low Positive $(s_{C,t})$ to Low Negative $(P_{C,count})$ doing action Go Right $(a_c)$, to influence Cart Velocity $(Ch(D))$, because Pole Velocity and Cart Velocity $(R(D))$ are connected to the goal. Following this action, it will be possible to obtain a final return of 100 $(d_{act})$, while following the counterfactual action will lead to a final return of 100 $(d_{count})$".

### 4.4.3   Mountain Car Environment

The second continuous environment analyzed in this set of experiments is Mountain Car, first developed in Andrew Moore's PhD thesis [170], and extensively presented in Appendix A.3 in the Electronic Supplementary Material. To solve this MDP, the DQN method is applied.

After delineating the environment, the BENEDICT methodology is applied to this scenario. Starting with the discretization procedure for learning the DAG of the BN. As previously explained in Cartpole, the unsupervised learning technique K-means is exploited in this problem and the optimal values for the number of clusters are selected using the elbow rule of thumb. The outcomes consisted of 5 groups for the position and 4 for the velocity. From this preliminary phase, the training dataset composed of 200 episodes is transformed into its discrete version and given as input to the NOTEARS-MDP algorithm. The graph discovered is portrayed in Fig. 4.18. In the testing process for assessing the quality of the estimation of the freshly generated BN, 100 episodes were run and the trajectories were compared to the prediction computed. The AUROC and accuracy achieved for each aspect forecasted are listed in Table 4.3. Even though in this case a discretization step was implemented, the calculations report overall satisfying results in terms of the efficiency in learning how the environment changes and in the action selection procedure of the DQN agent.

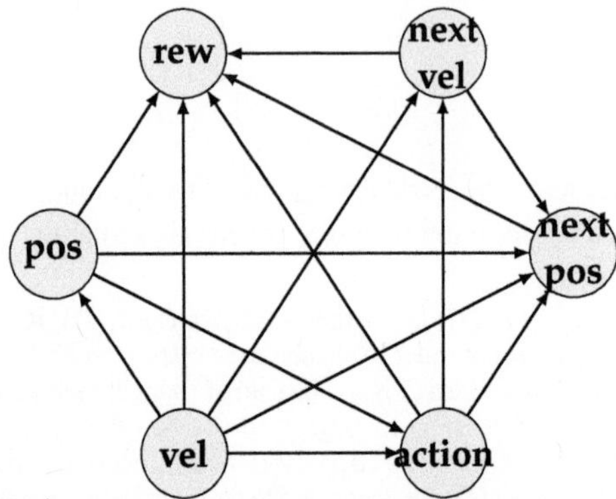


**Fig. 4.18** Structure of the BN for the Mountain Car environment. The names of the nodes represent the following features: "pos" is the position of the car, "vel" is the velocity of the car, "action" is the action, "next pos" is the next position of the car, "next vel" is the next velocity of the car and "rew" is the reward

**Table 4.3** AUROC and accuracy for the prediction of each future feature in the Mountain Car environment

| Parameter | Action | Position | Velocity |
| --- | --- | --- | --- |
| AUROC | 0.948 | 0.994 | 0.990 |
| Accuracy | 0.859 | 0.952 | 0.980 |

In the training of the distal GRU networks, it was chosen to select as relevant information the distal state representing the maximum height that was reached using the energy stored until that time step. Moreover, the reward function of the environment is modified in order to make the prediction more challenging. In particular, a piecewise function was used for increasing the speed of learning:

$$r(s_{t+1}) = \begin{cases} 100 & s_{t+1} \geq 0.6, \\ (1 + s_{t+1})^2 & 0 \leq s_{t+1} < 0.6, \\ 0 & s_{t+1} < 0, \end{cases} \tag{4.82}$$

where in case of successfully completing the task, the agent earns 100, and if it stays in the positive part of the map, corresponding to the uphill, it will receive a reward that increases with the x-position. In all the remaining situations, the reward is indeed set to 0. So, this new reward function encourages the agent to drive to the further part on the right. A remark about the starting position is necessary: since only in the positive values of the position the agent will receive a positive reward, the starting positions have to be on the negative side of the map, as done by setting them uniformly at random between $[-0.6, -0.4]$. The information required for the training of the GRU networks is then generated through the solution of the problem for 100 episodes with the new settings. From the plot of the losses during the training and test phases shown in Fig. 4.19, it can be observed that the computation of the distal position is straightforward. This outcome is expected given the linearity of the equations dealing with the calculation of the positions and velocities. Conversely, the prediction of the final return seems to present some instabilities, probably caused by the piecewise reward function. The focus point of this issue can be related to the different instants evaluated that could provide small biases depending on the exact moment when the position is calculated. In fact, by moving the position by a few decimals, it will produce a major change provoked by the squared exponent in the reward function definition. Nevertheless, it is interesting to notice that the GRU networks are capable of improving their estimation with an increasing number of

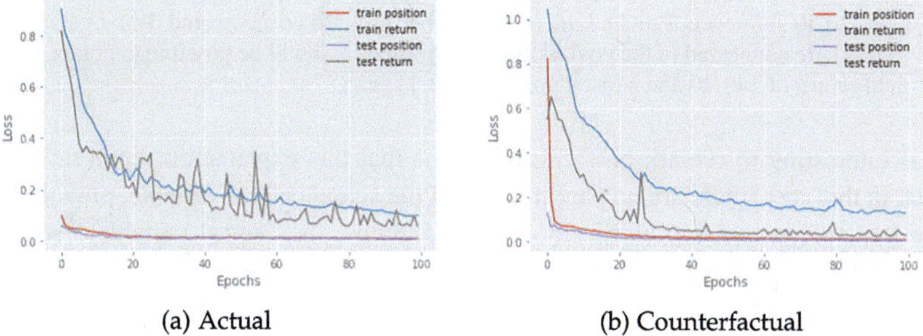

(a) Actual            (b) Counterfactual

**Fig. 4.19** Training and test losses for the Actual and Counterfactual RNNs in the Mountain Car environment

epochs. Therefore, with a larger dataset and extending the total amount of time for the learning process better results can be achieved.

In conclusion for this experimental part of Mountain Car, two practical examples with explanations for the actual and counterfactual situations are reported.

Let $s_t = $ (Left, Low Negative), where the first component denotes the position and, the second one, the velocity. The actual action chosen by DQN is $a_t = $ Go Left. From this data, the BN predicted that the consequent state values are:

$$P_{act} = (\text{Left, Negative}), \tag{4.83}$$

indicating that the central actual variable is the velocity. By looking at the DAG in Fig. 4.18, it is noticeable that the velocity has no parent node, meaning that the causal chain will start with this feature and then move to the child node represented by the position. Lastly, both the aforementioned vertexes have to be considered as reward nodes since they are connected to the reward feature in the graph. From the distal prediction, the results achieved are:

$$d_{act} = (148.20, 0.34), \tag{4.84}$$

where the first element denotes the final return obtained and the second one indicates the maximum x-position reached with that energy. Given this information, the semantic explanation to the question "Why Go Left?" can be constructed using Definition 4.9 as follows:

"Since Velocity is Low Negative $(s_{D,t})$, it is desirable to do action Go Left $(a_t)$ in order to change the Velocity $(D)$ from Low Negative $(s_{D,t})$ to Negative $(P_{D,act})$ to influence Position $(Ch(D))$, because Velocity and Position $(R(D))$ are connected to the goal. Following this action, it will be possible to obtain a final return of 148.20 and a distal position of 0.34 $(d_{act})$".

It is interesting to remark on the special case that this explanation represents. In fact, in this case, there are no parent nodes of the actual central variable, provoking the structure to not introduce any causes. Thereby, the central variable, and its instantiation, become the real starting point of the causal chain.

This specific scenario is also faced in the counterfactual instance. Let $\bar{a}_t = $ Go Right be the counterfactual action. Then, the BN outcome using this action as input together with the state instantiation is:

$$P_{count} = \texttt{(Left, Low-Positive)},\tag{4.85}$$

where the velocity is recognized as the counterfactual central variable. As already seen in the Cartpole example, the counterfactual and actual central variables are the same, but their predicted values are different. The motivation of the actual action is generated from this diversity. The counterfactual GRU network, looking at the memory stored for the episode so far, predicted as possible counterfactual distal information:

$$d_{count} = (143.34, 0.28).\tag{4.86}$$

Through this set of information, it is possible to thoroughly describe the reasons for "Why not $\texttt{Go Right}$?":

"Since $\texttt{Velocity}$ is $\texttt{Low Negative}$ ($s_{D,t}$), it is more desirable to do action $\texttt{Go Left}$ ($a_t$) in order to change the $\texttt{Velocity}$ ($D$) from $\texttt{Low Negative}$ ($s_{D,t}$) to $\texttt{Negative}$ ($P_{D,act}$) instead of changing the $\texttt{Velocity}$ ($C$) from $\texttt{Low Negative}$ ($s_{C,t}$) to $\texttt{Low Positive}$ ($P_{C,count}$) doing action $\texttt{Go Right}$ ($a_c$), to influence $\texttt{Position}$ ($Ch(D)$) because $\texttt{Velocity}$ and $\texttt{Position}$ ($R(D)$) are connected to the goal. Following this action, it will be possible to obtain a final return of 148.20 and a distal position of 0.34 ($d_{act}$), while following the counterfactual action will lead to a final return of 143.34 and a distal position of 0.28 ($d_{count}$)".

This final explanation concludes the testing sections, which provided the basic results and practical applications of the BENEDICT methodology that was previously introduced in theoretical terms. In the remaining part of the computational results, the comparison with the literature is discussed, outlining the major improvement that the proposed approach brought.

### 4.4.4  Discussion of the Results

By comparing the methodology proposed with the one suggested by Madumal et al. [148, 149], multiple observations can be noticed. The first problematic aspect relative to the approach in [148, 149] was regarding the discovery of the correct causal graph. In fact, in their case, the fundamental requirement for the creation of the action influence models is the possibility of using an expert in the field to handcraft the causal structure. In the BENEDICT method instead, this need is neglected since the DAG for the BN and the consequent probabilities can be directly learned from the data. Moreover, with the proper pruning, it is possible to take advantage of the characteristics of the MDPs in order to generate more reliable causal models.

Furthermore, with this specific choice, also the number of overall components used in the procedure can be reduced. In fact, through the exploitation of the BN modelling characteristics, not only the causal relationship between the variables can be derived but also the next time step information can be predicted. The accuracy of this approach has been compared in a benchmark evaluation study consisting of evaluating the predictions of 100 episodes in the three environments already presented in the previous sections: Taxi, Cartpole and Mountain car. The BN results are listed in Table 4.4, where the previous performances of the approximators used by Madumal et al. [148, 149] are also included. As can be noticed, in two out of three environments the forecasting qualities of the BN outperforms the other ML approaches. The only MDP where the BN are not achieving the best accuracy is Cartpole, where some limitations provoked by the discretization have already been discussed before.

**Table 4.4** Comparison of the accuracy (%) in prediction obtained with different methods: LR is the Linear Regression, DT is the Decision Tree, MLP is the Multi-Layer Perceptron, $DP$ is the Decision Policy Tree, $DP_n$ is the Decision Policy Tree without a fixed depth and BN is the Bayesian Network

| Env-RL | Size | Literature [148, 149] | | | | | Proposed Model |
|---|---|---|---|---|---|---|---|
| | | LR | DT | MLP | $DP$ | $DP_n$ | BN |
| Taxi-SARSA | 4/6 | 68.2 | 74.2 | 67.9 | 82.44 | 86.19 | **99.88** |
| Cartpole-PG | 4/2 | 83.8 | 81.6 | 86.0 | 96.83 | **97.10** | 78.13 |
| Mountain Car-DQN | 2/3 | 69.7 | 57.8 | 69.6 | 88.66 | 86.75 | **93.03** |

One issue that is still missing is the detection of the distal information. In the experimental test examined, this kind of data was smartly individuated through the general knowledge of the problem. However, this can not be generalized to any MDP, meaning that there can be issues related to the difficulty in the determination of what could be crucial for the later stages of the environment. Despite the lack of automatic detection of the distal information (which will be central in the next chapter), the overall procedure has shown remarkable results in terms of efficiency.

From the perspective of the explanations produced, the amount of data, that has been schematically selected and proposed to the user, is greater than the one used in the action influence model. This strategy may lead to understanding better how the RL agents are selecting particular actions instead of others. On the other hand, presenting longer sentences can also affect the clarity of the message that the answer is bringing. Thereby, in the next section, the attention will be steered towards the human evaluation study to finally comprehend if the BENEDICT pipeline can not only produce effective computational results but also increase the understandability of the RL policies.

## 4.5    Human Validation Study

The first question that has to be addressed before diving into the human evaluation study is: if an XAI method is provided to a user, how can be measured whether the explanation generated works and the user achieves a deeper understanding of the outcomes obtained from AIs? Basically, it is necessary to identify a set of metrics that can estimate how "good" the XAI method is. However, the definition of "goodness" has to be strictly related to objective concepts. To this end, the seminal work of Hoffman et al. [103] contributes to recognising the fundamental characteristics that discriminate the quality of an explanation: whether the users are satisfied by the explanations, how much the understanding of the AI methodology is increased by the explanations and whether the human testers' trust and reliance is improved [103].

All these traits have to be asserted first by the AI researchers who have developed the XAI approach. In this way, a methodology that can satisfy a "goodness" checklist for, at least, the experts-creators can be already designed. In the second step, a test on the satisfaction grade of independent judges has to be asked. In fact, while an explanation can be deemed good enough by some experts, it may be inadequate for some users [103]. By these means, the claims of the researchers can be checked and quantified. To this end, multiple key attributes must be determined from psychological literature on explanation [38, 173, 174]. The fundamental traits to be analyzed are understandability, the feeling of satisfaction, sufficiency of detail, completeness, reliability (safeness), accuracy in predictions (predictability), and trustworthiness [103]. These characteristics, which have already been defined in Chapter 3, are evaluated through a Likert scale [135] between 1 and 5, where 1 represent the worst grade and 5 the best. Through these measurements, it is then possible to directly obtain an objective estimation of the "goodness" for each of the aforementioned qualities.

After this brief introduction on the topic of validating explanations for AI methods, it is now possible to turn the focus on the proper user study carried out using BENEDICT as a central model.

### 4.5.1   Experimental Settings and Statistical Methodologies

In the user validation study that is described next, the aim is to verify the hypothesis that the explanations created through the BENEDICT method provide a better understanding of RL agents than the state-of-the-art causal approaches. To this end, a comprehensive comparative analysis is conducted. First, the MDP tackled in this experimentation has to be introduced: the Taxi problem. Specifically, two instances of the Taxi environment are presented to the tested users, one for each "Why" or "Why not" question. Under these circumstances, the explanations taken into consideration during this analysis, other than the one from BENEDICT, were the causal distal explanations generated by the action influence model discussed in Sect. 4.2.1 for both the actual and counterfactual situations and the benchmark explanation (template 1) suggested by Khan et al. [123] only for the "Why" question. The choice of the former is justified by the similarity in both the methodologies and the composition of the answers. The latter explanation individuated for this comparison was selected from the previous human study performed by Madumal et al. [149]. The inclusion of this established approach is reasoned by the possibility of contrasting the results that will be produced during this evaluation with the one already presented in the literature [149]. Moreover, it has been considered also the "None" explanation, which reports just the selected action, as a baseline for recognizing the usefulness of the explanations.

Fixed the general settings of the problem and scope of the experiment, the central object of the investigation is described: the survey adopted. As already stated, the main source for estimating the "goodness" of explanations from the AI process is provided by the questionnaires of Hoffman et al. [103]. Therefore, the survey adopted is based on the same concepts that were mentioned in the previous section. In detail, the first part of the questionnaire delivered to the testers consisted of an exhaustive description of the Taxi problem (to make them experts in this environment) and the overall tasks that they will have to face throughout the analysis. After the introductory part, an instantiation of the problem is shown and the relative validation questions are listed for each kind of explanation. In each of these blocks, the seven qualities of the explanations are graded using the Likert scale, as previously described. Additionally to the classical grading system used for giving scores to the attributes, preference questions and true and false statements are included at the end of each "Why"/"Why not" block of evaluations. Thereby, the users' thoughts about the different methodologies can be categorized in order to discover possible misunderstandings or improvements for the statement structures.

This survey has been undertaken anonymously by 30 students, self-selected at the University of Bundeswehr Munich and University of Camerino from different backgrounds (from the departments of Computer Science, Mathematics and Physics).

Thus, the audience surveyed consisted of people who had already knowledge of the RL field but also non-experts in data science and AI. By these means, it is possible to have a general overview of the sentiment gathered [64]. The data was collected in both physical and virtual manner through the application of online interfaces.

The statistical pipeline employed is shown in Fig. 4.20. One of the major phases in statistical studies is the verification of the fundamental hypothesis for which the tests are valid. In this case, the crucial assumptions, that have to be checked before deciding which statistical methodology to apply, are the independence of the evaluations, their normality and the homogeneity of the variances. The former requirement can be directly assumed to hold due to the differences in the subjects marking each explanation. For the remaining two conditions, specific tests have to be executed. Respectively, it was used the Shapiro-Wilk test [213] for the normality check and the Leveine's test [134] for the analysis of the variances. The results of the computation of these tests implied that the assumptions were not satisfied, hence a non-parametric approach had to be applied. For additional information about this preliminary step, are reported the details of each analysis in Appendix B.6 in the Electronic Supplementary Material.

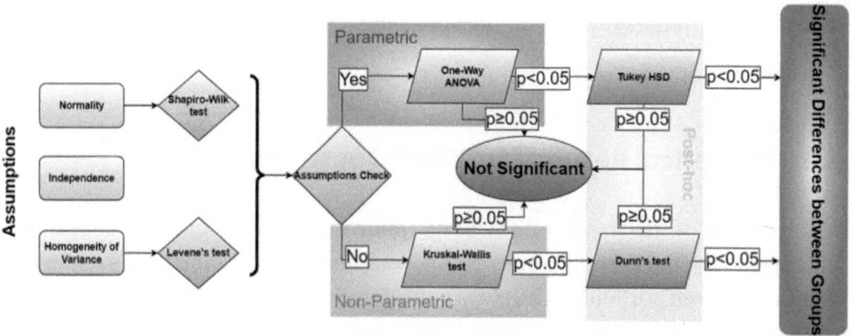


**Fig. 4.20** Statistical methodologies applied for the human evaluation study

Through the determination of the set of holding hypotheses, the correct path can be followed. The first part of the proper statistical pipeline consists in proving the existence of differences (in terms of rank sums) between the diverse groups from the data gathered. This particular evidence can be highlighted through the Kruskal-Wallis test [128], which represents the non-parametric version of the One-way ANOVA approach, which is widely misused in these contexts [100]. Indeed, the latter method can be utilized only when the normality assumption is satisfied

[259]. After applying the Kruskal-Wallis test, it is then possible to move to a post-hoc approach in the case of rejecting the null hypothesis consisting of having all the same central tendency in the samples. In fact, the Kruskal-Wallis test does not provide any information on which group is presenting a different tendency. Therefore, it is necessary to apply a pairwise comparison test like the Dunn's test [68]. This methodology is less powerful than the classical Tukey's HSD (honestly significant difference) test for a larger number of comparisons. Despite this pitfall, it is fundamental to recall that the normality and homogeneity of variance conditions are not verified, implying that it is not possible to use Tukey's test [242]. Hence, the reasons for adopting the Dunn's test are clarified. Additionally, the Bonferroni correction [34] was applied for the $p$-values obtained in the Dunn's test in order to solve the multiple comparisons problem [66] (by correcting the family-wise error rate following the analysis of the variance in the post-hoc process). In such a manner, the Bonferroni correction tends to be more conservative by reducing the threshold for rejecting the null hypothesis [86]. This two-phase procedure is repeated for each attribute that has been studied and for each specific question that has been addressed.

The results collected through the discussed statistical pipeline are shown in the subsequent sections. In particular, the description of the outcomes is divided into the actual and counterfactual blocks in order to reduce confusion. Furthermore, the results of the different preference questions are also included in the examination below.

## 4.5.2   Statistical Results for "Why" Questions

The description of the results is structured following the same order of appearance in the survey. So, the first outcomes outlined are the scores attained by each explanation in the seven particular attributes considered. Before evaluating any possible statistical significance, a visualization of the marks given by the human participants is shown in Fig. 4.21. In these boxplots, it is possible to directly identify where the means (denoted as diamonds) and the medians (thick line) of each group of explanations are positioned. At first glance, the highlighted aspect in these plots is the overall best performance achieved by the proposed methodology (Milani et al. [159]). Moreover, it can be observed that the marks received for Khan et al. [123] and Madumal et al. [149] are similar, providing further confirmation of the results that were reported in [149] where both the approaches have been evaluated as equivalent. Although through this visualization it seems clear that there are possible differences between the proposed answer to the "Why" question and the state-of-the-art statements, it is required to verify if there are any statistically significant results. Hence, the previously introduced statistical pipeline is applied.

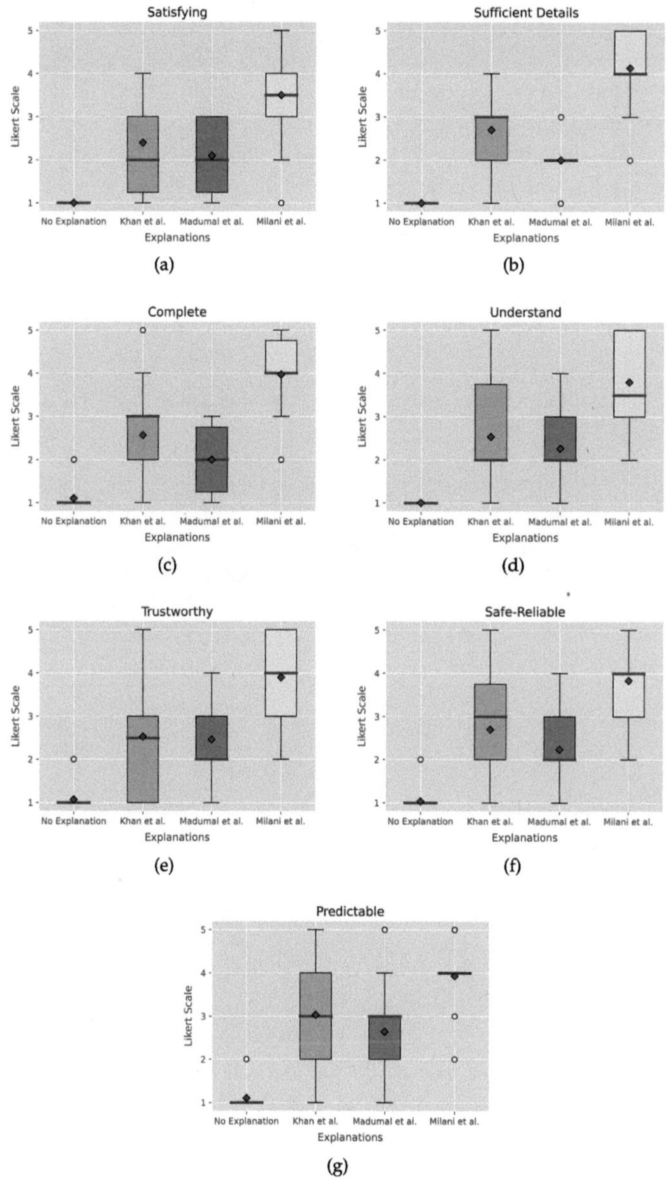

**Fig. 4.21** Box plots of the marks associated with each attribute of the explanations for the "Why" question

Then, the Kruskal-Wallis test is performed and the resulting $p$-values and $F$-values for each corresponding attribute are listed in Table 4.5. From the computed values, it is possible to confirm the existence of significant differences between diverse explanation groups. In fact, not only the $p$-values are lower than the classical threshold (0.05), but also the $F$-values are higher than the critical value (2.69), i.e., the between-group variance is large or the within-group variance is small, or both the conditions hold. Moreover, the significant distance of the F- and $p$-values to the critical values demonstrates the existence of large gaps in the different groups analyzed. The next step consists of applying a post-hoc test to individuate which specific groups significantly differ from each other.

**Table 4.5** Table of the $F$-values and $p$-values obtained in the Kruskal-Wallis test for the "Why" question explanations

|                    | $F$-value  | $p$-value     |
| ------------------ | ---------- | ------------- |
| Understand         | 72.960691  | 9.910749e-16  |
| Satisfying         | 62.098886  | 2.092725e-13  |
| Sufficient Details | 88.564855  | 4.454306e-19  |
| Complete           | 74.628244  | 4.352924e-16  |
| Trustworthy        | 65.861863  | 3.280842e-14  |
| Predictable        | 72.218146  | 1.429504e-15  |
| Safe-Reliable      | 72.177046  | 1.458781e-15  |

To this end, the Dunn's test is carried out and the corrected $p$-values for each comparison and quality are reported in Table 4.6. The first trivial conclusion that can be drawn concerns the better scores achieved by the explanations if compared to the "None Explanation". This obvious result testifies how these explanatory statements can help increase the user's knowledge about the RL agent method. Nevertheless, it is fundamental to recognize which is the best explanation for this task, since different methodologies can give a diverse level of satisfaction in the comprehension of the RL policies [194]. A second remark follows from the comparison between the group of the action influence model and the relevant variables: no significant differences between the statements of Khan et al. [123] and Madumal et al. [149] have been found in any attribute. Thereby, the visual observations identified previously are statistically validated, together with the results shown in [149]. Regarding the answer created by the proposed BENEDICT pipeline, statistically significant differences with the other causal and relevant variables approaches are discovered.

These discrepancies are well-defined in all the cases but one: in the comparison with Khan et al. [123], the p-value associated with the predictable feature is not lower than the threshold of 0.05; implying that there are no statistically satisfying conditions for stating that there is a difference.

**Table 4.6** Table of the $p$-values obtained from Dunn's test for the comparison of the "Why" question explanations. The names of the respective groups are shortened as follows: N = No Explanation,  R = Khan et al. [123],  C = Madumal et al. [149],  B = Milani et al. [159]

|  | R - N | C - N | B - N | C - R | B - R | B - C |
|---|---|---|---|---|---|---|
| Understand | <0.001 | <0.001 | <0.001 | 1.000 | 0.004 | <0.001 |
| Satisfying | <0.001 | <0.001 | <0.001 | 1.000 | 0.020 | 0.001 |
| Sufficient Details | <0.001 | 0.001 | <0.001 | 0.327 | 0.002 | <0.001 |
| Complete | <0.001 | 0.008 | <0.001 | 0.746 | 0.001 | <0.001 |
| Trustworthy | <0.001 | <0.001 | <0.001 | 1.000 | 0.001 | 0.002 |
| Predictable | <0.001 | <0.001 | <0.001 | 1.000 | 0.057 | 0.001 |
| Safe-Reliable | <0.001 | <0.001 | <0.001 | 1.000 | 0.009 | <0.001 |

These outcomes are also supported by the evaluation of the preferences given by the participants to the survey, as shown in Fig. 4.22. In these questions, it was asked to recognize the best explanation in terms of being convincing, trustworthy and easy to understand. In particular, the suggested explanation achieved the best overall preferences in each characteristic, reaching the 83.3% for "Convincing", 83.3% for "Trustworthy" and 60% for "Easy to Understand" of the consensuses.

In addition, these findings are endorsed by the answers given by the tested users in the true and false statement block. In this part of the questionnaire, the BENEDICT explanations are directly evaluated positively or negatively for five different qualities: being easy to understand, providing more information, being accurate, requiring more information and being transparent. The results are shown in Fig. 4.23. Even in this validation analysis, the proposed methodology is considered positive in the majority of the aspects: 96.7% of the surveyed participants positively validated the accuracy and transparency of the statement written, 86.7% confirmed that the explanation reported more information than the other and 76.7% indicated it as easy to understand. The only balanced situation is found in the necessity of additional information. Therefore, quite half of the population of this human study would like to have more data depicted in the explanations in order to gain deeper insights into the RL action-selection process.

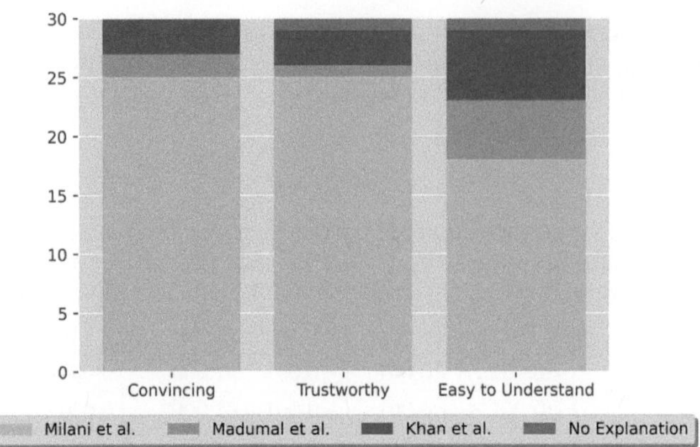


**Fig. 4.22** Bar chart of the preferences relative to specific qualities assigned by the participants to the explanations of the "Why" question

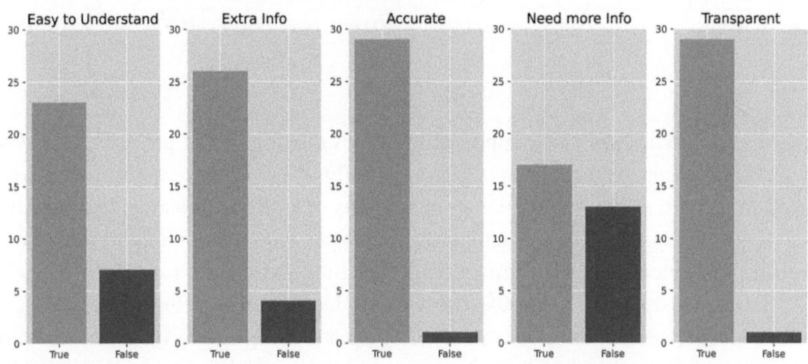


**Fig. 4.23** Bar chart representing the results of the true and false statements results indicating the characteristics of the proposed explanation for the "Why" question

Based on this statistical study, it is possible to conclude that the explanations for "Why" questions generated through the BENEDICT methodology are the best performing in terms of different attributes and general preferences.

### 4.5.3   Statistical Results for "Why not" Questions

As done in the previous section for the "Why" question, the data gathered from the human evaluation study in the counterfactual tests are also visualized in box plots in Fig. 4.24. For each of the seven characteristics analyzed, the presence of possible differences is not clear, except in the case of the comparisons with the "No Explanation" where the gap is evident. In these box plots, the elements that do not allow to distinctively recognize a ranking between the kind of explanations are the intersecting areas between the $25^{th}$ and $75^{th}$ percentile of the various groups. In fact, these shared portions can invalidate the qualitative depiction of relevant conclusions from these graphs. Therefore, a more detailed study has to be executed.

Thus, the Kruskal-Wallis test is carried out to find possible evidence of distinctions between the three methodologies studied. The resulting $F$-values and $p$-values computed through the aforementioned statistical method are indicated in Table 4.7. Despite this counterfactual scenario having slightly worse values (greater $p$-values and small $F$-values) than in the actual case, significant differences are also detected here for each tested quality. Taking into account the remark about the small deviations achieved in the calculated statistical values for the "Why not" results, a smaller amount of significant differences between groups will be discovered in the post-hoc analysis outlined below.

The observation stated above is confirmed using the multiple comparisons of Dunn's test. As previously specified, the $p$-values reported in Table 4.8 have been adjusted through the use of the Bonferroni correction. The trivial outcomes of the test verify the assumption that the explanations are regarded as helpful, similar to the "Why" question context. Turning the focus towards the more interesting comparisons between the proposed methodology and the one from the literature, it can be noted that the $p$-values are substantially greater than in the previous section analysis. These signs highlight the smaller variation in the perception between the distal explanation created by Madumal et al. [148] and the one discussed here [159]. Only for three attributes there is a significant discrepancy that is validated statistically; explicitly, for Sufficient Details (0.001), Complete (0.015), and Safe-Reliable (0.020). Nothing can be objectively stated about the other qualities since the $p$-values associated with these comparisons are above the classical threshold.

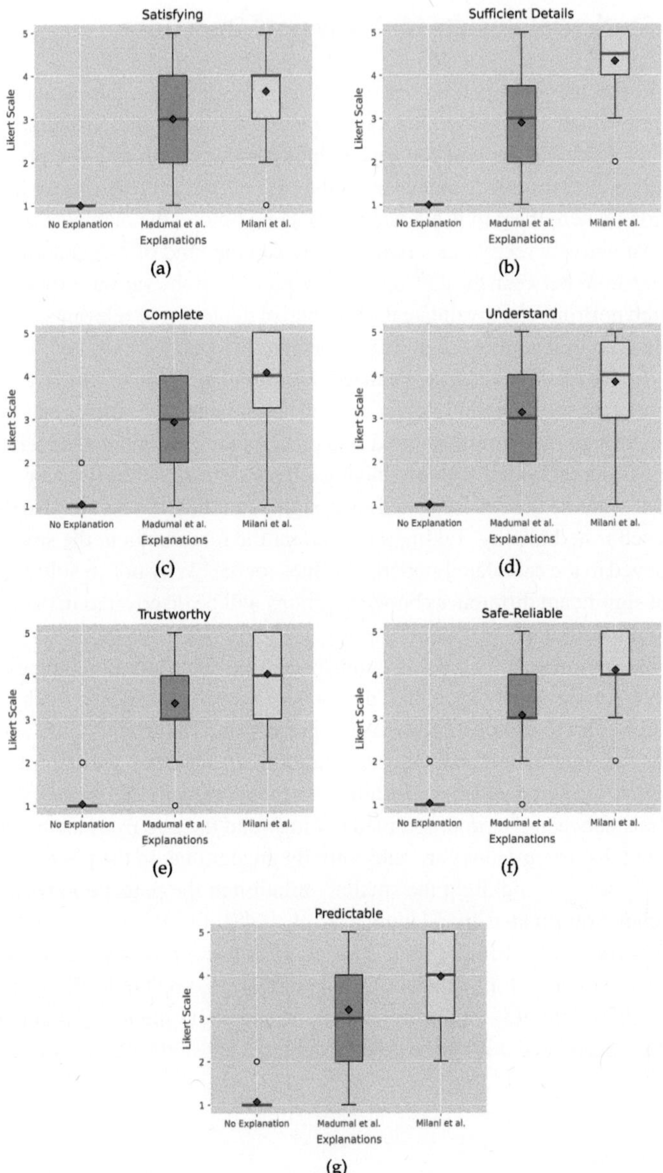



**Fig. 4.24** Box plots of the marks associated with each attribute of the explanations for the "Why not" question

**Table 4.7** Table of the $p$-values obtained from the Kruskal-Wallis test for the comparison of the "Why no" question explanations

|                    | F-value    | p-value       |
|--------------------|------------|---------------|
| Understand         | 64.481313  | 9.955449e-15  |
| Satisfying         | 61.631290  | 4.139381e-14  |
| Sufficient Details | 71.454554  | 3.046777e-16  |
| Complete           | 64.447933  | 1.012300e-14  |
| Trustworthy        | 63.839649  | 1.372134e-14  |
| Predictable        | 60.428756  | 7.552008e-14  |
| Safe-Reliable      | 67.743373  | 1.948558e-15  |

**Table 4.8** Table of the $p$-values obtained from Dunn's test for the comparison of the "Why not" question explanations. The names of the respective groups are shortened as follows: N = No Explanation,  C = Madumal et al. [148],  B = Milani et al. [159]

|                    | C - N    | B - N    | B - C  |
|--------------------|----------|----------|--------|
| Understand         | <0.001   | <0.001   | 0.226  |
| Satisfying         | <0.001   | <0.001   | 0.376  |
| Sufficient Details | <0.001   | <0.001   | 0.001  |
| Complete           | <0.001   | <0.001   | 0.015  |
| Trustworthy        | <0.001   | <0.001   | 0.240  |
| Predictable        | <0.001   | <0.001   | 0.192  |
| Safe-Reliable      | <0.001   | <0.001   | 0.020  |

Moreover, the explanation generated through the BENEDICT methodology has been selected as the best in being "Convincing" (73.3%) and "Trustworthy" (86.7%) according to the results shown in Fig. 4.25. Therefore, the superiority of the proposed method over the action influence model can be remarked. However, it is essential to notice a critical aspect involved in both explanations. In fact, the statements used in answering the counterfactual question were judged as the easiest to understand with an equal partition of the surveyed population.

This particular finding suggests that both the explanations (the proposed ones and the ones generated by te action influence graphs) present similar levels of complexity. Nonetheless, if double-checked with the outcomes from the true and false statements in Fig.  4.26, then it is evident that the explanations considered are not always clear. In fact, only 56.7% of the tested users recognized the BN-derived explanation

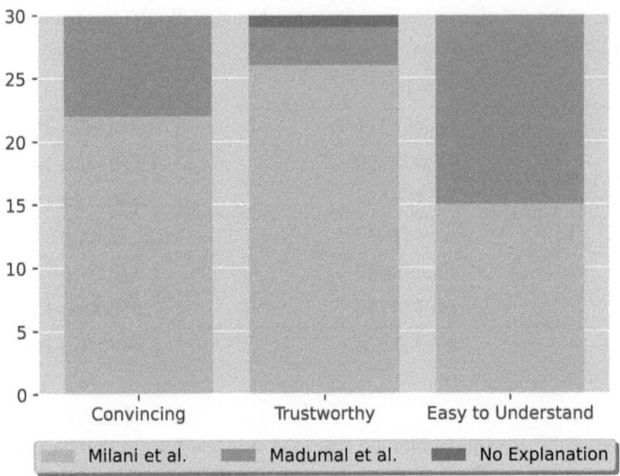



**Fig. 4.25** Bar chart of the preferences relative to specific qualities assigned by the participants to the explanations of the "Why not" question

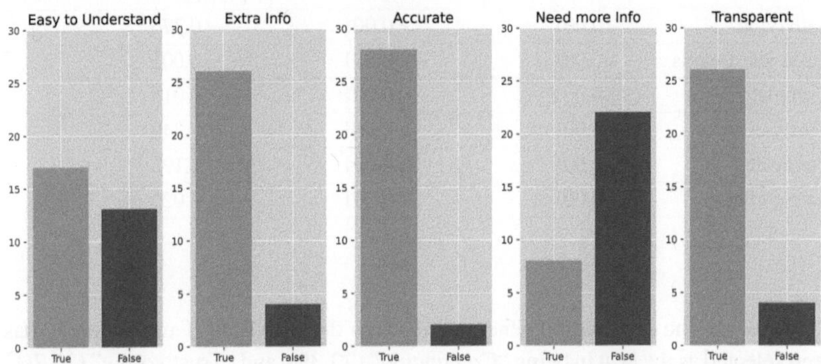



**Fig. 4.26** Bar chart representing the results of the true and false statements results indicating the characteristics of the proposed explanation for the "Why not" question

as easy to comprehend, implying that the clarity of the semantic structure can be improved. This result is caused by the greater amount of data employed for justifying the choice of action in the counterfactual situation. Indeed, the testers stated that it is not necessary to increase the amount of information used in 76.7% of the answers. This result can be also justified by two additional factors: the users first saw the answers to the "Why" questions and completed the survey for that part, and only in a second moment they focused on the counterfactual scenarios. Thus, they have already processed a significant amount of information before entering the "Why not" section of the questionnaire. Moreover, due to the simplicity of the Taxi environment, deeper specifics are not required to improve the understanding of the optimal policies.

Furthermore, the studied population indicated how the proposed methodology could also present more insights into the RL policy (86.7%) without invalidating its accuracy (93.3%) and transparency (86.7%). Therefore, the overall analysis proves how the BENEDICT procedure can generate highly informative explanations for counterfactual questions, but, apparently, at a small cost in terms of ease of understandability. Hence, the superiority of the novel approach is also confirmed for the "Why not" questions.

### 4.5.4  Discussion of the Human Study Results

To summarize the results achieved using the BENEDICT methodology in the improvement of the explanations generated, statistically significant differences have been found during the human evaluation test for both cases, i.e., in the actual and counterfactual scenarios. Specifically in the former case, the answer created by the proposed methodology for the "Why" question scored way higher marks with respect to the literature baselines, and objectively being tested as the best explanations for all the seven considered qualities. Alternatively, in the latter situation, the results obtained presented a smaller number of evident differences. This remark can be related also to the outcome of the comparisons for each attribute, where the amount of statistically relevant discrepancies between the groups dropped to three. Despite having a smaller set of characteristics where the suggested approach performed better than the state-of-the-art methods, the users from the human study assigned, on average, higher grades to the explanations created by BENEDICT. The aforementioned result is confirmed by Fig. 4.27, where the radar charts presenting the mean values of the scores assigned by the tester population in both the actual (Fig. 4.27b) and counterfactual (Fig. 4.27a) problems are shown.

Moreover, from the preferences and false and true statements, it has been found that the users favoured the freshly presented answers instead of the classical ones. In fact, an aspect that was widely appreciated was the inclusion of a larger set of information that could bring them more insights into the hidden mechanism of choices of the RL agent. In particular, this increment of the data used in the explanations was possible through the introduction of the counterfactual distal information. While in Madumal et al. [148] the focus is only on the actual version of the distal action, in

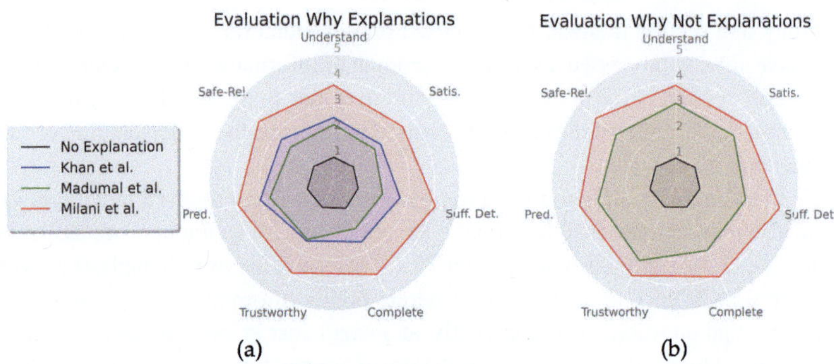


**Fig. 4.27** Radar charts of the mean scores obtained for each quality of the explanations for: (a) "Why" and (b) "Why not" questions

the proposed pipeline it was possible to generate sharper explanations through the extension of the concept to states and specific values such as the final return, other than also broadening it to the "Why not" question. Especially through the comparison of the later stages situations represented by the distal actual and counterfactual information, it can be improved the sense of trust in the actions selected by the RL algorithm.

After this discussion about the results reported from the human evaluation study of the explanations generated with multiple XRL methodologies, in the next section, the chapter will be concluded by answering the associated RQ.

## 4.6 Answering Research Question 1

In this chapter, the attention was focused on the creation of causal explanations for model-free RL agents. The end goal of the methodology developed was to answer the RQ1 presented in this thesis:

**RQ1**: How to automatically generate causal explanations that can be human-validated?

Through the adoption of a particularly structured BN, it is possible to learn the causal relationships between the variables in the MDP and also predict the future changes derived from the decisions of the RL algorithm in the environment. The specific DAGs for the BNs have been created using an improved version of the

NOTEARS algorithm able to exploit the main characteristics of the MDPs, leading to recognizing the connections more accurately. From this point onward, all the processes related to the production of the causal explanation are obviously automatic. Nevertheless, it is fundamental to satisfy also the requirement of being human-validated, meaning that users' satisfaction with the suggested answers to "Why" and "Why not" questions achieve higher levels of understanding than the state-of-the-art approaches. To this end, the concept of distal action has been extended to include also more general information, such as the distal states and distal values like the final return. To complete this task, two GRU networks specialized in the forecasting of respectively the actual and counterfactual distal information have been trained. By combining both the BN, delineating the causal component, and the two GRU networks, depicting the distal components, the full BENEDICT pipeline is outlined. With this precise procedure, the causal explanations can be automatically created and the human users are convinced of the outcomes obtained.

Despite improving the perception qualities of the explanation, the addition of the distal components raises a major problem: due to the absence of reliable metrics that can discriminate the distal state and actions, it is necessary to handpick them, in order to generate the datasets for the GRU networks learning process. Therefore, the achievements reached by the updated methods, leading to the limitation (and elimination) of the human interventions in the cycle, are partially invalidated. For this reason, the RQ2 is stated and the intention in the next chapter is to delineate an answer.

# Recognition of Important States

**5**

The focus of this chapter is on the development of importance metrics for the detection of the distal states. After delineating the characteristics of a particular situation, it is possible to recognize the associated action to discriminate the distal actions. Nevertheless, it is fundamental to clearly define the properties that are strictly linked to the distal attribute. In particular, all the states that can be visited only after performing a chain of precise actions must be relevant for the completion of the problem. Thus, their appearance in trajectories may increase the final return achieved at the end of the episode, or at least modify it significantly.

Moreover, these particular states can be considered to be bottlenecks due to the small amount of possible combinations of actions that will lead to their activation. For example, in the Taxi environment, the states where the taxi is in the same location as the passenger are frequently visited positions by various trajectories. Hence, there is a restriction of the solution in that particular state. Therefore, it is required to identify the characteristic properties of a bottleneck state and also the state features affecting directly the outcome of the problem.

However, no metrics from the literature attain good performances in detecting these types of properties [158]. The importance measures, that have been studied so far, can be split into Q-value-based and graph-based. In the former class, these methodologies argue that the importance of a state has to be linked accordingly to the state-action value function estimations [7, 8]. Despite providing a classical prediction of the states that play critical roles in determining the highest final return, nothing can be said about the positioning of the selected states in the overall environment structure. Indeed, no information about the possibility of them being a

**Supplementary Information** The online version contains supplementary material available at https://doi.org/10.1007/978-3-658-50495-3_5.


R. Milani, *Advanced Automation for Comprehensible Causal Explanations of Reinforcement Learning Agents*,
https://doi.org/10.1007/978-3-658-50495-3_5

bottleneck is analyzed. Moreover, by applying these measures, the states that are closer to the final goal in a specific environment will gain higher importance values, due to the greater Q-values associated with them.

On the other hand, in the graph-based group, the main idea adopted is to represent the visited states as nodes of a transition network [195]. In this setting, classical centrality measures can be utilized to spot important states [223]. In this case, the focus is turned to the analysis of the explored network of states. Thus, the most influential states exploit concepts relative to shortest path and connectivity properties, which can be directly related to the bottlenecks [224]. Nonetheless, as is discussed in the literature review reported in the following pages of this chapter, no fruitful application of these approaches has been tested in contexts where the rewards were involved as well. That implies the analytical tools employed did not care about the final return that would have been gained by passing through a particularly central state.

Starting from the actual state-of-the-art metrics adopted for this task, this chapter will introduce measures that will be able to couple together the aforementioned properties to identify the proper distal states. Specifically, new methodologies are introduced for both groups of metrics. The results for the Q-value-based approaches are reported in *Milani et al.* [158], while for the graph-based approaches, are described the methodologies reported in *Milani et al.* (submitted) [160].

The derived analysis is structured as in the previous methodological chapter, where, in the first place, the background information relative to Graph Theory and associated centrality measures in MDPs are outlined. Then, it is possible to dive deeper into the multiple advancements introduced in the two aforementioned papers. Finally, a brief discussion of the overall results and possible future implications are highlighted.

## 5.1 Preliminaries

The background information, helpful to fully comprehend the approaches introduced in the following pages, delves into the Graph Theory notions. Although some of the basic concepts relative to the simple graphs and DAGs have been encountered in the chapter before, here more advanced definitions are reported. Particularly interesting are the multilayer networks and the centrality measures associated with both multigraphs and multilayer graphs. These theoretical topics will be essential for understanding the graph-based approaches, while, the MDPs ideas are the crucial ones for understanding the Q-value-based methods.

### 5.1.1 Multigraphs and Multilayer Networks

In this section, the description is centered on advanced graph structures. Although some simple concepts have been used in the previous chapters, e.g., the simple (directed) graphs and DAGs, in Appendix C.1 in the Electronic Supplementary Material the narrative will briefly cover these terms to clearly define the notation. Subsequently, they are extended here coherently to the more advanced formulations of the multigraphs and multilayer networks. A short remark consists of avoiding the repetition of the attribute *direct* for all the graphs that are described below. Then, all of them have to be considered with edges pointing in a specific direction (node).

The first fundamental definition presented is the *weighted edge-coloured multigraph*. A *weighted edge-coloured multigraph* $G_M = (V, \hat{E}, \hat{W})$, is a weighted graph where the conditions of simplicity are eliminated, meaning that this framework allows for the presence of parallel edges and self-loops. For this reason, the newly used edge set is denoted as $\hat{E} \subseteq V \times V \times C$, where $C$ indicates the set of the labels (or colours) [125]. Due to the presence of this additional information relative to the parallel edges, the creation of the weighted adjacency matrix is complicated. Classically, the weighted adjacency matrix for the multigraphs can be obtained through the summation of all the parallel edges, i.e., $\hat{w}_{ij} = \sum_{c \in C} w_{ij}^c$ with $w_{ij}^c$ associated to $e_{ij}^c \in \hat{E}$, connoting the weight connected to the edge linking node $i$ to $j$ with colour $c$. So, the structure of the graph is simplified and the usual methodology for representing the weighted adjacency matrix can be applied.

Despite the adoption of parallel edges can generate issues in the representation of the multigraphs, they allow the introduction of the concept of the multilayer networks $M = (\mathcal{G}, \mathcal{E})$, where $\mathcal{G} = \{G^h = (V^h, E^h, W^h), \forall h = 1, \ldots, |L|\}$ is a family of weighted simple graphs, and $\mathcal{E} = \{(E^{h,k}, W^{h,k}) \mid E^{h,k} \subseteq V^h \times V^k, \forall h, k = 1, \ldots, |L|\}$ is the set of the inter-layer connections $E^{h,k}$ between the layers $h$ and $k$ with weights $W^{h,k}, \forall h, k = 1, \ldots, |L|$. It can be noticed how each label in $C$ can be associated with a layer. Thereby, it is possible to generate a weighted simple graph composed of all the vertices in the initial multigraph and with connections that show the same colours [27, 49]. Even though the aspects that highlight the relationships between multigraphs and multilayer networks are multiple, it is not relevant to the aim of this chapter to dig deeper. Actually, the only class of multilayer networks involved in this work is the *multiplex networks* one. Therefore, the focus is turned in this direction now.

The particularity of these graphs lies in the shared set of vertices for each layer. Moreover, a restriction to the inter-layer connections is introduced: the only links present are connecting the same vertices from different layers. Thereby, the mathematical formulation of the multiplex networks with inter-layer weights $\omega_I$ can be defined as $G^L = \{G^h = (V, E^h, W^h), \forall h = 1, \ldots, |L|\}$ [31]. It is helpful to point out that there are no dependencies between $\omega_I$ and $W^h$, $\forall h = 1, \ldots, |L|$, since the first ones represent only the inter-layer weights and the latter the intra-layer weights. Furthermore, the notion of weighted adjacency matrix can be used in this context to summarize the information relative to the multiplex networks. In detail, the corresponding matrix, called *supra-adjacency matrix*, able to represent the multilayer graphs is as follows:

$$
B = \begin{bmatrix}
W^1 & W^{1,2} & \cdots & W^{1,|L|} \\
W^{2,1} & W^2 & \ddots & \vdots \\
\vdots & \ddots & \ddots & W^{|L|-1,|L|} \\
W^{|L|,1} & \cdots & W^{|L|,|L|-1} & W^{|L|}
\end{bmatrix},
\tag{5.1}
$$

where $W^h$, $\forall h = 1, \ldots, |L|$ denotes the weighted adjacency matrix of the graph $G^h$, and $W^{h,k}$, $\forall h, k = 1, \ldots, |L|$ indicates the weighted adjacency matrix associated to the inter-layer connections between graph $G^h$ and $G^k$. For the multiplex case, the supra-adjacency matrix will simplify as follows:

$$
B = \begin{bmatrix}
W^1 & \omega_I I_{|V|} & \cdots & \omega_I I_{|V|} \\
\omega_I I_{|V|} & W^2 & \ddots & \vdots \\
\vdots & \ddots & \ddots & \omega_I I_{|V|} \\
\omega_I I_{|V|} & \cdots & \omega_I I_{|V|} & W^{|L|}
\end{bmatrix},
\tag{5.2}
$$

where $I_{|V|}$ expresses the identity matrix of order $|V|$.

Specific metrics rely deeply upon these particular mathematical representations of multigraphs and multiplexes. Thus, it is indispensable to fix these concepts in the best manner possible. Following the thread, in the next section, the centrality measures adopted throughout this thesis are introduced with the theoretical assumptions required to ensure their existence.

### 5.1.2   Centrality Measures

A centrality measure is defined as a function $C^G : V \rightarrow \mathbb{R}$, where $C(v)$ is the centrality of node $v \in V$ in the graph $G$. The aim of centrality measures consists of depicting the importance of single entities in a network by formally evaluating a mathematical measure. One of the major issues in the network analytics field is the absence of unanimity on what the concept of centrality is [75]. Due to its numerous definitions, no agreement on the proper methodology to measure it can be pointed out. Consequently, a wide range of centrality measures for different sets of graphs have been developed. In this section, the discussion is restricted to the ones that have been already considered in the literature and the ones that will be significant for the proposed approaches.

The first centrality measure addressed is the *betweenness centrality*. This particular metric assesses the importance of each node by calculating the ratio between the number of shortest paths passing through that vertex and the overall total number of shortest paths [12]. Mathematically, it can be defined as:

$$C_B^G(v) = \sum_{s \neq v \neq t} \frac{\sigma_{st}(v)}{\sigma_{st}} \quad \forall v \in V, \tag{5.3}$$

where $\sigma_{st}$ is the total number of shortest paths from node $s$ to $t$ and $\sigma_{st}(v)$ denotes the number of shortest paths from node $s$ to $t$ passing through node $v$ [75]. The definition of betweenness centrality is well posed, under the assumptions of the absence of negative cycles in the graph, i.e., a cycle where the sum of the weights of the selected edges is less than 0. Alternatively, a negative cycle can be exploited to obtain an infinitely negative gain leading to the non-existence of the shortest path. Although this is the only requirement that has to be satisfied, the general formulation of betweenness can be directly exported to multiple kinds of graphs. The adaption of this mathematical concept is straightforward for both the multigraphs and the multilayer networks, especially for the multiplexes. In all of these situations, the main idea remains the same: the most important nodes are the ones that have been used for reaching the destination in a shorter amount of time (cost). Thereby, the bottlenecks can be individuated quickly through this kind of metric.

A similar measure is the *closeness centrality*, which uses the concept of shortest path distances to discriminate the bottleneck vertices. In this case, it is not counting the number of shortest paths passing through a node, but the relative distance to each of the other nodes [21, 22]. By applying this idea, a higher importance is assigned to the vertices that are closer to all the remaining ones. Formally, it is defined as follows:

$$C_C^G(v) = \frac{1}{\sum_{s \in V} d(s, v)} \quad \forall v \in V, \tag{5.4}$$

where $d(s, v)$ represents the (shortest path) distance between the nodes $s$ and $v$ [207]. From the formulation reported in Equation (5.4), it can be noticed how critical the concept of distance is. While for the unweighted simple graphs, if the usual length of an edge is fixed at 1, it is possible to have always a non-negative distance (the equality to 0 is used for the distance between a node and itself), it can not be done for the weighted graphs. In the latter scenario, it is vital to impose the non-negativity of each edge weight. Alternatively, if this assumption is not verified it may happen that the distances are cancelling with each other, thus, obtaining a value close to 0 in the denominator or finding a negative centrality value; which is not permitted. Hence, the measure would lose consistency.

Additionally, it is not possible to straightforwardly modify the edges and maintain their meaning and ordering. For example, by adopting the absolute value of the distances, the magnitude of importance of each edge is reversed for the negative weight edges, leading to being identified as equally relevant as the positive weight edges. Another simple idea that could be tried is to translate all the weights by the minimum value. However, also in this scenario, the structure of the shortest path is completely distorted, losing its initial interpretation.

A different problem arises when computing the distances between two nodes that are not connected by any paths. In this case, the general rule-of-thumb suggests indicating the shortest path distance as equal to $\infty$. By doing so, the final centrality measure will be (informally) equal to 0. Therefore, it is essential to apply this measure to only *strongly connected graphs*, i.e., graphs where there exists at least one path between each node of the graph. Nevertheless, this assumption is strong since it is not always true, especially for directed graphs. Thus, a formulation derived from the same main idea and not necessitating strong connectivity has to be examined.

A metric satisfying the above properties has been theorized by Beauchamp [23] and it is reported below:

$$C_{CB}^{G}(v) = \sum_{s \in V \setminus \{v\}} \frac{1}{d(s, v)} \quad \forall v \in V, \tag{5.5}$$

with the convention that if there are no paths between two nodes, then the contribution of that term is 0. With this definition, it has completely avoided the problem of strong connectivity. Despite the elimination of this condition, the requirement of the non-negativity of the weights has to hold.

Also in this case, the transition from simple graphs to multigraphs and multiplexes is direct. In fact, all these concepts can be translated into the corresponding structures bearing in mind that the non-negativity condition has to be maintained at all the edges. So, it has to be also satisfied for each of the parallel edges and inter-layer connections.

For the last centrality measure discussed in this section, the *eigenvector centrality*, a specific methodology for the different forms of graphs has to be studied. To evaluate this measure, the different shapes adjacency matrices that have been described in the previous section are utilized. To clarify all the steps that lead to the generalizations of the eigenvector centrality to the advanced graph structures, it is first necessary to introduce the classical formulations.

For a simple graph, the *eigenvector centrality* can be calculated by finding the eigenvalues and eigenvectors of the weighted adjacency matrix $A$ associated with the graph. Specifically, by evaluating the eigenvector of the largest positive eigenvalue of matrix $A$, it is possible to recover the different importance values for this metric [33]. In fact, the *eigenvector centrality* of a node $v \in V$ is represented by the $v$ component of the eigenvector of the largest positive eigenvalue. This measure has the ability to recognize the vertices that are more well-connected to the rest of the graph. In fact, the eigenvector centrality of a node is proportional to the sum of the centralities of its neighbors. With the addition of the weights, the importance of each node can be scaled depending on how many relevant vertices are connected with them.

In order to prove the existence of this metric, the conditions under which there is a positive eigenvalue for the associated adjacency matrix have been studied. This classical result has been proven by Perron [189] and Frobenius et al. [77] considering different initial assumptions. Before jumping to the theorem stating the existence conditions for the eigenvector centrality, it is necessary to introduce a mathematical concept that is used in the following theorem: the *irreducibility* of a matrix.

**Definition 5.1**  *A matrix $A \in \mathbb{R}^{n \times n}$ is defined as reducible if there exists a permutation matrix $P \in \mathbb{R}^{n \times n}$ such that:*

$$C = PAP^T = \begin{bmatrix} A_{11} & A_{12} \\ \mathbf{0} & A_{22} \end{bmatrix}, \tag{5.6}$$

*where $A_{11} \in \mathbb{R}^{r \times r}$, $A_{22} \in \mathbb{R}^{(n-r) \times (n-r)}$, $A_{12} \in \mathbb{R}^{r \times (n-r)}$ and $\mathbf{0} \in \mathbb{R}^{(n-r) \times r}$ is the zero matrix, with $0 < r < n$.*

*If a matrix $A \in \mathbb{R}^{n \times n}$ is not reducible, then it is defined as irreducible.*

For the definition above, it is helpful to remember that a permutation matrix $P \in \mathbb{R}^{n \times n}$ consists of a square binary matrix that has exactly one entry equal to 1 in each row and column and 0 in the others. Moreover, it is also useful to bear in mind that permutation matrices are orthogonal, i.e., $P^{-1} = P^T$. After this brief reminder of the characteristics of the permutation matrices and the introduction of the concept of irreducibility of a matrix, Perron-Frobenius Theorem can be stated:

**Theorem 5.1** (Perron-Frobenius). *Let $A \in \mathbb{R}^{n \times n}$ be a non-negative irreducible matrix with spectral radius $\rho(A) = \max\{\lambda_1, \ldots, \lambda_n\}$, where $|\lambda_1|, \ldots, |\lambda_n|$ are the eigenvalues of $A$, then:*

1. *$\rho(A) > 0$ and $\rho(A)$ is an eigenvalue of $A$ with multiplicity one;*
2. *$\exists \bar{x}, \bar{y} \in \mathbb{R}_{\geq}^n \setminus \{0\}$ s.t. $A\bar{x} = \rho(A)\bar{x}$ and $\bar{y}^T A = \rho(A)\bar{y}^T$, i.e., the left and right eigenvectors of $A$ associated with the eigenvalue $\rho(A)$ are both positive.*

For more details regarding the proof of the just introduced result, refer to Appendix C.2 in the Electronic Supplementary Material.

Based on the Perron-Frobenius Theorem (Theorem 5.1), the existence conditions for the eigenvector centrality measures can be proved. Moreover, the uniqueness of the eigenvector centrality can be ensured using claim 1. of the Perron-Frobenius Theorem (Theorem 5.1), and imposing that the vector is normalized. In this way, it is eliminated the multiplicative invariance of the eigenvectors. On the other hand, for the existence conditions, the adjacency matrix must be non-negative and, in particular, irreducible. While the first assumption is clear, i.e., all the entries of the adjacency matrix have to be positive or equal to 0, for the second hypothesis, it is complex to grasp a practical property that can characterize the irreducibility.

Finally, it can be proved that irreducibility and strong connectivity are equivalent. Thereby, a pragmatic condition can be applied directly to the graph.

**Proposition 5.1** *A matrix $A \in \mathbb{R}^{n \times n}$ is irreducible if and only if the associated graph $G$ is strongly connected.*

**Proof.** Let $A \in \mathbb{R}^{n \times n}$ be an irreducible matrix. Suppose by contradiction that $G = (V, E)$ is not strongly connected, i.e., $\exists v_i, v_j \in V$, such that there is no paths between $v_i$ and $v_j$. Therefore, the set of vertices in $G$ can be divided into two non-empty sets: $S_1$ which contains all the nodes connected to $v_j$ and $S_2$ which, instead, groups the nodes that have no connection with $v_j$. No connections between the nodes in $S_1$ and $S_2$ are present, otherwise, the vertices in $S_2$ would also be in $S_1$. Now, assuming that $r$ and $n - r$ are the cardinality of $S_1$ and $S_2$ respectively, the indexes can be permuted as follows: first write the nodes $v_1, v_2, \ldots, v_r \in S_1$ and then $v_{r+1}, v_{r+2}, \ldots, v_n \in S_2$, obtaining the matrix $C$. In this way, it can noticed that $c_{hk} = 0$ for all $h = r + 1, r + 2, \ldots, n$ and $k = 1, 2, \ldots, n$. However, $A$ is irreducible, resulting in a contradiction.

The opposite direction of the statement can be proved in a similar manner.   $\square$

With this operative property, the question of whether the eigenvector centrality measure exists or not can be addressed directly by looking at the graph.

The extension of this metric to the weighted simple graphs can be straightforward by calculating the eigenvector of the weighted adjacency matrix. Nevertheless, it is essential to bear in mind that the weights must be non-negative to satisfy the hypothesis of the Perron-Frobenius Theorem (Theorem 5.1). Obviously, this approach can be applied also to multigraphs, with the precaution of summing up all the parallel edges, in order to have just a single weight for each connection between two nodes. The classical methodology consists of using the weighted adjacency matrix where all the weights corresponding to the parallel edges are summed up [177]. By applying this preliminary step, the entries of the resulting matrix are still non-negative (if the initial ones were non-negative).

The situation gets more complicated in the scenario of calculating the eigenvector centrality in the multiplexes. The first idea that comes to mind is to directly use each adjacency matrix for the different layer graphs to generate one matrix. Hereby, given a multiplex $G^L$ with inter-layer weights $\omega_I$, the following matrix can be computed:

$$\tilde{W} = \sum_{h=1}^{|L|} W^h \tag{5.7}$$

where $W^h$ is the weighted adjacency matrix of the $h$-th layer graph. This approach evaluates the eigenvector centrality for $\tilde{W}$, and it is called the *uniform centrality* of the multiplex $(G^L, \omega_I)$ [226]. Noticeably, the inter-layer connections are irrelevant in this formulation. Furthermore, given a multiplex network $G^L$ with inter-layer weights $\omega_I$ with $G^L = \{G^h = (V, E^h, W^h), \forall h = 1, \ldots, |L|\}$, the projected network of $G^L$ is defined as $\tilde{G} = (V, \tilde{E}, \tilde{W})$ where $\tilde{E} = \cup_{h=1}^{|L|} E^h$. Therefore, it can be observed that the computed $\tilde{W}$ can represent the weighted adjacency matrix of the projected network $\tilde{G}$. Hence, the Perron-Frobenius Theorem (Theorem 5.1) can be applied to ensure the existence of the eigenvector centrality.

**Theorem 5.2**  ([226]) *Given a multiplex network $G^L$ with inter-layer weights $\omega_I$, if its projected graph $\tilde{G}$ is strongly connected and the resulting weighted adjacency matrix $\tilde{W}$ is non-negative, then the uniform centrality metric exists and is unique.*

A different approach for the evaluation of the eigenvector centrality for the multilayer networks considers the adoption of the fourth-order tensor representation as suggested by De Domenico et al. [57, 58]. The authors defined the multilayer adjacency tensor $\mathcal{B} = (\mathcal{B}_{i,j}^{h,k})_{i,j=1,\ldots,|V|}^{h,k=1,\ldots,|L|}$, where $\mathcal{B}_{i,j}^{h,k}$ indicates the weight associated with the edge connecting the node $i$ in layer $h$ to the node $j$ in layer $k$. From this tensor and the application of the variables matrix $F \in vR^{|V| \times |L|}$, it can be written a system of equations similar to the classical eigenvector-eigenvalue formulation:

$$\sum_{j=1}^{|V|}\sum_{k=1}^{|L|} \mathcal{B}_{i,j}^{h,k} F_j^k = \lambda_1 F_i^h, \quad \forall i = 1, \ldots, |V|, \ \forall h = 1, \ldots, |L|, \tag{5.8}$$

and the *eigenvector versatility* of node $i$ in layer $h$ is defined as the component $F_i^h$. Interestingly, Equation (5.8) can be simply calculated through the exploitation of the supra-adjacency matrix $B \in \mathbb{R}^{|V||L| \times |V||L|}$ associated with the multilayer adjacency tensor $\mathcal{B}$. Then, if the spectral radius of $B$ is evaluated and denoted as $\lambda_B$, it can be proved that $\lambda_B = \lambda_1$ [57]. Moreover, when computed the eigenvector $x_B \in \mathbb{R}^{|V||L|}$ of $B$ linked to $\lambda_B$ by solving $Bx_B = \lambda_B x_B$, it can be shown that $x_B = vec(F)$, meaning that it is the vectorization of the tensor $F$ satisfying Equation (5.8) [58].

Therefore, for the existence conditions of the eigenvector versatility, it is possible to apply the Perron-Frobenius Theorem (Theorem 5.1). Nevertheless, the assumptions required are always limiting the practical applicability of these measures. Thus, the last method reported is just assuming weaker hypotheses in order to satisfy the existence and uniqueness condition for the eigenvector metric for multiplexes.

The idea is based on a third-order representation of the multiplexes presented by Tudisco et al. [241]. Let $\mathcal{A} = (\mathcal{A}_{i,j}^k)_{i,j=1,\ldots,|V|}^{k=1,\ldots,|L|}$ be the non-negative adjacency tensor of a weighted multiplex $G_L = \{G^h = (V^h, E^h, W^h), \forall h = 1, \ldots, |L|\}$, where $\mathcal{A}_{i,j}^h$ is the weight associated with edge $e_{i,j}^h$ connecting vertices $i$ and $j$ in the layer $h$. A remark can be stated: all the eigenvector centrality measures described in the context of multiplexes are not influenced by the inter-layer weights due to their obvious presence for each node.

After identifying this third-order representation, it is necessary to state the system of equations related to the calculation of the eigenvalues:

$$\begin{cases} \sum_{j=1}^{|V|} \sum_{h=1}^{|L|} \mathcal{A}_{i,j}^h x_j t^h = \mu(x_i)^\alpha & \forall i = 1, \ldots, |V|, \\ \sum_{i=1}^{|V|} \sum_{j=1}^{|V|} \mathcal{A}_{i,j}^h x_i x_j = \lambda(t^h)^\beta & \forall h = 1, \ldots, |L|, \end{cases} \tag{5.9}$$

for some positive scalars $\mu$ and $\lambda$. Then, assuming that the positive constants $\alpha, \beta > 0$ are such that $\frac{2}{\beta} < \alpha - 1$ for any vertex $v \in V$ and layer $h \in L$, it is possible to define the $f$-*node eigenvector centrality* of node $v$ as $C_f^{G_L}(v) = x_v$ and the $f-layer$ *eigenvector centrality* of layer $h$ as $C_{f,L}^{G_L}(h) = t^h$, where $(x, t)$ is the unique non-negative eigenvector of a multi-homogeneous function $f$, i.e., the unique solution of the system of nonlinear Equations (5.9). Particularly, the aforementioned mapping $f$ can be defined as $f = (f_1, f_2) : \mathbb{R}_{\geq}^{|V|} \times \mathbb{R}_{\geq}^{|L|} \to \mathbb{R}_{\geq}^{|V|} \times \mathbb{R}_{\geq}^{|L|}$ with:

$$f_1(x, t)_i = \left( \sum_{j=1}^{|V|} \sum_{h=1}^{|L|} \mathcal{A}_{i,j}^h x_j t^h \right)^{\frac{1}{\alpha}}, \quad f_2(x, t)_h = \left( \sum_{i=1}^{|V|} \sum_{j=1}^{|V|} \mathcal{A}_{i,j}^h x_i x_j \right)^{\frac{1}{\beta}},$$

$$(5.10)$$

for all $i = 1, \ldots, |V|$, $h = 1, \ldots, |L|$. This definition can be obtained from the left-hand side of Equation (5.9), and by eliminating the exponent of the variables on the right-hand side. In this way, it is generalized the interpretation of the eigenvector centrality in the multiplexes.

In fact, the $f-$node eigenvector centrality is able to estimate the centrality of a node through the connections of the corresponding vertices in each layer. Furthermore, the $f-$layer eigenvector centrality can compute the importance of each layer depending on the strength of the edges in them. Thereby, a more in-depth analysis is provided through these measurements.

Before moving to the calculation of these metrics, it is remarkable to notice that if there exists an index $i$ such that $\mathcal{A}_{i,j}^h = 0$, $\forall j = 1, \ldots, |V|, \forall h = 1, \ldots, |L|$, then $x_i = 0$. In the same way for the layer $h$ for which $\mathcal{A}_{i,j}^h = 0$, $\forall i, j = 1, \ldots, |V|$ implies $t^h = 0$. So, depending on the non-zero patterns of $\mathcal{A}$, the solutions of Equation (5.9) have some entries equal to 0.

Focusing now on the practical characteristics of this methodology, it is helpful to first discuss a pragmatic approach for the calculation of the $f$-eigenvector metric. Consequently, the assumption for the existence and uniqueness can be explained straightforwardly. The following set of pairs of vectors are introduced:

$$\mathcal{P}_{\mathcal{A}}^{|V| \times |L|} = \left\{ (x, t) \in \mathbb{R}_{\geq}^{|V|} \times \mathbb{R}_{\geq}^{|L|} \mid \begin{array}{ll} x_i \sim \sum_{j,h} \mathcal{A}_{i,j}^h, & \forall i = 1, \ldots, |V| \\ t^h \sim \sum_{i,j} \mathcal{A}_{i,j}^h, & \forall h = 1, \ldots, |L| \end{array} \right\},$$

$$(5.11)$$

where, given two non-negative numbers $x, y \geq 0$, the relation $x \sim y$ denotes the existence of a positive constant $C > 0$ such that $x = Cy$. Thereby, the non-trivial solutions of Equation (5.9) are in the set $\mathcal{P}_{\mathcal{A}}^{|V| \times |L|}$. In this case, the uniqueness can be proved only up to a scalar multiple: id $(x, t)$ is a solution of Equation (5.9), then also $(ax, bt)$ is a solution for any $a, b > 0$. Obviously, the scalar $\mu$ and $\lambda$ are different in the second scenario. So, in order to ensure uniqueness it is possible to restrict the aforementioned space as follows:

$$\mathcal{S}_{\mathcal{A}}^{|V| \times |L|} = \{(x, t) \in \mathcal{P}_{\mathcal{A}}^{|V| \times |L|} \mid ||x||_1 = ||t||_1 = 1\}, \tag{5.12}$$

where the only vectors considered have to be normalized for each corresponding component. Then, the normalized version of the mapping f (defined in Equation (5.10)) can be obtained as $g : \mathcal{S}_{\mathcal{A}}^{|V| \times |L|} \to \mathcal{S}_{\mathcal{A}}^{|V| \times |L|}$ which can be explicitly written as:

$$g(x, t) = \left( \frac{f_1(x, t)}{||f_1(x, t)||_1}, \frac{f_2(x, t)}{||f_2(x, t)||_1} \right). \tag{5.13}$$

The crucial property of this map relies on the study of its fixed points: if $(x^*, t^*)$ is a solution of Equation (5.9), then it is a fixed point of $g$. Indeed, the relation $g(x^*, t^*) = (x^*, t^*)$ is Equation (5.9). Conversely, it can be shown that any fixed point of $g$ solves Equation (5.9). To formally prove this result, it has to be verified that $g$ is a contraction and then the Banach fixed point Theorem [122] has to be applied. From this, follows the next result:

**Theorem 5.3** ([241]) *Let $\mathcal{A} \in \mathbb{R}_{\geq}^{|V| \times |V| \times |L|}$ be a non-zero non-negative tensor and let $\alpha, \beta > 0$ be s.t. $\frac{2}{\beta} < \alpha - 1$. Then, Equation (5.9) has a unique normalised non-negative solution $(x, t)$.*

Moreover, it is possible to obtain the positiveness of solution $(x, t)$ by assuming two simple requirements:

1. $\forall h \in L$, $\exists i, j \in V$ s.t. $A_{i,j}^h > 0$, i.e., for each layer, there is at least one edge with a positive weight.
2. $\forall i \in V$, $\exists h \in L$ and $\exists j \in V$ s.t. $A_{i,j}^h > 0$, i.e., for each node, there is an edge with a positive weight in at least one layer.

These hypotheses ensure the positiveness of an eigenvector, which means that the requirements have been weakened from imposing the irreducibility/strong connectivity assumption. Thus, these metrics can be easily applied without completely modifying the structure of the graph analyzed, and still maintaining their consistency.

In the next pages, the central topic that is outlined involves a deep analysis of the literature methodologies for the recognition of important states in MDPs. Specifically, attention is turned to the direction of the importance metrics derived using the Q-values and the centrality measures that have been already delineated.

## 5.2   Specific Literature: Importance Metrics in XRL

The relevance of efficiently detecting important states in MDPs has increased in the last years due to the numerous applications that can be derived from that: from increasing the understanding of the RL agents' choices to improving the training of the same algorithm [7, 9, 212]. Together with the importance of this task, it increased also the number of approaches dealing with this problem [158]. Therefore, a classification of the possible methodologies applied has to be structured. To this end, due to the extensive number of methods that have been developed in the literature, the following overview of the state-of-the-art procedures is restricted to the analysis of the importance measures employed. In detail, the importance metrics adopted two main concepts for recognizing the fundamental states: the Q-values (Sect. 2.2) or the centrality measures (Sect. 5.1.2). From the former, it can be generated the Q-value-based group of measures which includes all the methodologies exploiting the knowledge learned in the form of the state-action value function. On the other hand, the latter concept can be used in conjunction with the notion of *reconstructed transition graph*, namely, the graph composed of the states visited during the learning phase of an RL agent. In this way, the different network analytic measures can be directly utilized in these graphs for creating a ranking between the different nodes (denoting the states).

Both the aforementioned classes of metrics are discussed in depth in the successive pages.

### 5.2.1   Q-value-based Metrics

As already pointed out, the metrics described in this first group rely on the Q-value function learned during the training. Nonetheless, this aspect already provides a critical issue: depending on the kind of RL agent that has been trained, the values found may be different, thus, implying that the associated metrics will strictly relate to the specific agent. Hence, the measures that are examined in this class are agent-dependent.

The chronologically first importance measure that has been used in the MDPs and RL context is the *advising metric* [239]. The mathematical formulation of the advising importance metric is the following:

$$I_A(s) = \max_{a \in \mathbb{A}} Q(s, a) - \min_{a \in \mathbb{A}} Q(s, a), \quad \forall s \in \mathbb{S}, \tag{5.14}$$

which indicates the difference between the maximum and the minimum Q-value, fixed the state that has to be evaluated. This particular measure was applied in a completely different context (student-teacher framework) by [47]. Only in a second moment, Torrey and Taylor [239] adapted it in the XRL context. Through this measurement, it is indeed possible to estimate the overall quality of taking the correct action in contrast to the worst-case scenario. With this gimmick, the states where the optimal choice presents a larger gap with the poorest outcome are ranked in higher positions. Despite performing decently in the interpretation of policies, this measurement can be applied to determining the relevant states for improving the learning process of the RL agents too [7].

Instead of basing the importance on the difference between the highest and lowest values of the state-action value function, the following value can also be computed:

$$I_{AG}(s) = Q(s, a^*) - \max_{a \in \mathbb{A}^*} Q(s, a), \quad \forall s \in \mathbb{S}, \tag{5.15}$$

where $a^* = \arg\max_{a \in \mathbb{A}} Q(s, a)$ is the action maximizing the Q-value function in $s$ and $\mathbb{A}^* = \mathbb{A} \setminus \{a^*\}$ represent the set of actions except the optimal one. This special metric is called the *action gap importance* [24, 108], since it provides a measure of how good is a state based on the performances of the corresponding best action to select with respect to the second best. In this situation, the situations that are highlighted are the ones where one action is the only optimal decision. All the remaining options will indeed provide a smaller final return. Therefore, while the advising measure determines the overall criticality of a state based on the decision of the optimal action, the action gap estimates the importance focusing only on the quality of the best action. So, the advising importance can be used mostly for discriminating possible risky states while the action gap indicates the bottlenecks achieving the highest final return.

On the same thread of the previous two metrics, a natural formulation can be derived: the *mean value importance* [106]. The main idea that has been used by the authors proposing this metric is to evaluate the importance of a state depending on how much better the performance of the optimal action is compared to a random policy. Mathematically, this concept can be formulated as follows:

$$I_M(s) = \max_{a \in \mathbb{A}} Q(s, a) - \frac{\sum_{a \in \mathbb{A}} Q(s, a)}{|\mathbb{A}|}, \quad \forall s \in \mathbb{S}, \tag{5.16}$$

where $|\mathbb{A}|$ denotes the number of actions. It has to be emphasized that the settings where this particular metric is well-defined are the finite-time finite-space MDPs.

Lately, more advanced metrics have been derived from the combination of classical Q-values and innovative processes. An example in this direction is the *lazy measure* [114]. This specific measure can be seen as an extension of the previous mean value importance. Indeed, with the lazy measure, the randomness can be modelled through the adoption of a (default) lazy policy $\bar{\pi}$ which can be used in safe occasions. Formally, the lazy measure is defined below:

$$I_L(s) = \max_{a \in \mathbb{A}} Q(s, a) - \mathbb{E}_{\bar{\pi}}\big[Q(s, a)\big], \quad \forall s \in \mathbb{S}, \tag{5.17}$$

which computes the gap between the value of the best action and the expected return achieved following the lazy policy $\bar{\pi}$. Thereby, the discriminated states are the ones where it is fundamental to take a specific action since the default action is not able to complete the problem satisfactorily.

A metric that instead couples together the ML approaches with the Q-values is proposed by Cheng et al. [44], where the authors introduced a state-mask to recognize the states that will entail higher changes in the final return if a non-optimal action is performed. Practically, in order to create this approximation, a second RL agent (PPO in [44]) is trained to choose between two possible actions: 0 if the action that will be decided by the agent is fundamental to achieving the best outcome and 1 if a random action different from the optimal one can provide decent results. In this way, the state-mask can generate a continuous value that can be associated with the (reversed) importance of a state.

So far the metrics described are always comparing the maximum Q-value associated with a specific state to a second value, as summarised in Table 5.1. However, it is possible to evaluate the quality of a state together with the action selected as done by Wang et al. [255] without the usage of the maximum Q-value. In this work, the perturbation derived from small changes in the state is used as a measure of importance. Hence, the settings where this metric can be applied are the continuous space MDPs. The metric can be mathematically described as follows:

$$I_P(s, a) = \frac{|Q(s + \delta, a) - Q(s, a)|}{\delta}, \quad \forall s \in \mathbb{S}, \ \forall a \in \mathbb{A}, \tag{5.18}$$

where $\delta$ represents the vector of perturbations.

**Table 5.1** Q-value-based methods using the formula $\max_{a \in \mathbb{A}} Q(s, a) - \xi(s)$, where $\xi(s)$ is the interchangeable second term

| Reference | Name | Second Term $\xi(s)$ |
|---|---|---|
| [239] | Advising metric | $\min_{a \in \mathbb{A}} Q(s, a)$ |
| [108] | Action gap | $\max_{a \in \mathbb{A}^*} Q(s, a)$ |
| [106] | Mean value importance | $\frac{\sum_{a \in \mathbb{A}} Q(s,a)}{|\mathbb{A}|}$ |
| [114] | Lazy gap | $\mathbb{E}_{\tilde{\pi}}\big[Q(s, a)\big]$ |

Despite the last three procedures achieving positive effects in increasing the interpretability of the state selection process, they also involve the use of complex tools which have to be well-calibrated beforehand. Therefore, the focus of the approaches that have been introduced will be on the first simplistic measures.

## 5.2.2   Graph-based Metrics

In this section, the analysis of the literature delves into the approaches related to the usage of graph representations. In detail, the sub-field majorly addressed is automatic skill acquisition, where the relevant actions have to be recognized [155]. In this context, it is fundamental to be able to detect first the important states where particular operations have to be carried out to solve the MDP. To this aim, the core concept applied in almost all the methodologies reported here consists of the *reconstructed transition graph*. In this kind of graph, the visited states are depicted as nodes and the actions connecting two states are matched with the edges connecting the two vertices. Hence, the building process of the reconstructed transition graph can be performed simultaneously with the training of the RL agent: every time a new state is reached, a node will be included in the graph and connected to the previous vertices through the action that leads to its discovery.

However, as already pointed out, this representation has not always been used. Indeed, the pioneering paper in the context of finding relevant states [62] did not involve the aforementioned graph. In this work, Digney decided to discern the important states according to the number of visits that occurred and the gradient of the reinforcement signals obtained. At first glance, this could appear as a graph-agnostic approach due to the absence of any network-related notions. Nevertheless, though not clearly stated in the original paper [62], the main idea resembles the node centrality definitions for complex networks [32]. Indeed, multiple similarities can be spotted when comparing the visitations with the betweenness, closeness, and other centrality measures [75].

For this reason, the consequent improvement that was developed in this field was to adapt the well-known centrality measures to the scenario of MDPs through the application of the reconstructed transition graphs. The first to accomplish this shift from the analysis of the number of visitations and the actual employment of centrality measures were Şimşek et al. [225]. Moreover, the same authors started tackling this problem following the same analysis of visits of states [222], and only in a second moment switched to the examination of the reconstructed transition graphs [223, 225]. Specifically, they focused their investigation on the application of betweenness centrality metrics in directed and unweighted reconstructed transition graphs [221, 223, 224]. The results achieved consisted of the discovery of non-trivial sub-goals in simple grid-world environments.

Although the usage of the betweenness is clearly related to the detection of fundamental states, multiple other options that can solve this task are available for the same networks settings [155]. For example, a wide range of clustering techniques have been implemented in the MDPs to determine the different ranking groups between the states [70, 150, 214, 225]. In this way, not only the critical states can be precisely aggregated, but also the set of sub-goals can be delineated through a local partitioning of the graph [225]. More recently, an approach adopting the Louvain algorithm [30] for finding partitions that maximize the modularity of a graph has been implemented in the automatic skill acquisition field, resulting in positively defining a hierarchical order of sub-goals [70]. Other clustering techniques applied in these settings were using topological or value information [150], immune clustering algorithms [214] and strongly connected components analysis [120].

As shown in [120], the interesting states can also play a fundamental role in maintaining the connectivity of the reconstructed transition graph. For this reason, the methodologies that can associate higher relevance to states that present great connectivity have been exploited for the reconstructed transition graphs. An example is the evaluation of the classical eigenvector centrality for the detection of highly connected nodes [235]. Additionally, special metrics have been created. The first that has been implemented is the connection graph stability proposed by Rad et al. [195], where the authors define a centrality measure combining the length of the shortest path (for each starting and ending node) passing on a particular state and the total number of shortest paths linking the same two vertices. A natural extension to this methodology has been suggested by Moradi et al. [171], who included the importance of neighboring nodes to expand the knowledge derived from a local to a global perspective.

In all the previous procedures, the reconstructed transition graph was used as a directed and unweighted network. Thus, all the statistics regarding the reward function of the MDP are lost. To limit this loss, Shoeleh and Asadpour [217] merged the classical transition graph with the distance graph obtained from the rewards stored. Thereby, the final graph analyzed included weights for each edge [217, 218]. By using similar weighted networks, Menache et al. [154] built a Max Flow—Min Cut problem for detecting the bottleneck states.

After listing the major contributions relevant to discerning the important states, in the next section, the limitations of the state-of-the-art methodologies are discussed. From these observations, it will later be possible to define a clear path of research leading to the contributions reported in this thesis.

## 5.2.3 Limitations of Existing Measures

For all the metrics that have been described in the literature review above, it can be observed how no one tried to discriminate the exact distal states. One reason is related to the recent definition of this particular concept. Therefore, the combination of the properties achieved by those states has not been addressed together. The specifications of the distal states comprise a component relative to the satisfactory attained final return, and a part related to the fundamental chain of actions and states that have to be selected in order to reach the distal state. Hence, the latter relevance in the position of the trajectory and the topological state space can be straightforwardly attached to the concept of bottleneck. On the other hand, the importance of visiting a particular state can be evaluated through an accurate analysis of the rewards and based on the completion of the goal.

Despite all these attributes having been studied in separate manners, it is notable that neither the Q-value-based approaches nor the graph-based can be extended easily to include missing information. Indeed, for both classes, more advanced research is required to create a methodology that can reliably find the distal states.

In particular, for the Q-value-based procedures, a fundamental issue is related to their characteristic of being completely static, i.e., the importance values computed do not depend on the specific instant when the state is appearing and do not remember the number of visitations. In doing so, it is complex to address the bottlenecks present in a MDP using the pure Q-values. Therefore, it is needed to couple together the aforementioned aspects to deviate the importance measure towards the highly visited states that are also in the central part of the trajectories. The methodologies in this class already take into account the final return in the form of the state-action values, meaning that the positive outcome can be estimated.

On the other hand, the problem is reversed for the graph-based methods. In this class, the implementation that is hard to include concerns the approximation of the final return. While in the majority of the reported studies it has been possible to efficiently recognize bottlenecks and highly connected states, it was not necessary to know information about the rewards gained. Hence, in the graph-based group, the focus is only on the analysis of the topological characteristics. Moreover, just a few papers addressed the problem of extending the reconstructed transition graphs into weighted networks, but without producing optimal results.

The improvements proposed are discussed in the following sections, one for each group of metrics. In this way, the different enhancements that have been introduced to refine the process of detecting distal states can be clearly addressed, also specifying the assumptions required in each case.

## 5.3  Improving Q-value-based Metrics

The methodologies and the resulting outcomes obtained from the computational tests described in this section have already been published in Milani et al. [158]. The structure of the next subsections is as follows. Initially, a third complementary static Q-value-based metric, that can be directly associated with the two classical measures (advising and action gap), is derived. Then, a procedure to include the aforementioned visitation examinations is proposed. This is fundamental to identify the distal states. Lastly, the complete set of suggested methods have been evaluated using three discrete MDPs dealing with simple transportation tasks.

### 5.3.1  A Novel Metric: The Difference Measure

In the previous overview of importance metrics for the Q-value-based group, the two classical measures considered the difference between the maximum Q-value and the minimum (advising measurements [239]) and the second largest Q-value (action gap [24, 108]). It can be noticed how both these values are strictly addressing the importance of a state depending on the quality of the best-case scenario, i.e., when the optimal action is chosen. Hence, they are not able to discriminate high-risk situations that could terminate the episode with a small accumulated reward. However, a measure that can accomplish this goal can be crafted by avoiding to use the best action. Indeed, comparing the second-best Q-value to the minimum shows how critical a state is: when a large value is computed in this manner, then it identifies a risky situation given by the presence of a bad outcome for, at least, one action.

Formally, it can be defined as follows:

$$I_D(s) = \max_{a \in \mathbb{A}^*} Q(s, a) - \min_{a \in \mathbb{A}} Q(s, a), \quad \forall s \in \mathbb{S}. \tag{5.19}$$

More in detail, this measure can be also expressed as a combination of the advising and action gap. The following expression is obtained with a few algebraic steps:

$$I_D(s) = \max_{a \in \mathbb{A}^*} Q(s, a) - \min_{a \in \mathbb{A}} Q(s, a) \tag{5.20}$$

$$= \underbrace{\max_{a \in \mathbb{A}^*} Q(s, a) - \max_{a \in \mathbb{A}} Q(s, a)}_{-I_{AG}(s)} + \underbrace{\max_{a \in \mathbb{A}} Q(s, a) - \min_{a \in \mathbb{A}} Q(s, a)}_{I_A(s)} \tag{5.21}$$

$$= I_A(s) - I_{AG}(s) \tag{5.22}$$

For this reason, the novel metric introduced is named *Difference metric* since it can be deduced from the difference between the advising and action gap measurements.

By using this metric, it is possible to recognize states that present at least an action leading to a critical loss in the final return. Therefore, it can help in discerning high-risk situations. When comparing the results with one of the other two standard measures, a clearer overview of the scenario can be gained. For example, the comparisons of the estimations of the difference metric with the action gap can lead to spot occasions where all the actions different from the optimal ones are wrong. When the action gap is large but the difference measure is small, then the final return attains a larger profit only by following the best policy. With this combined observation, multiple criticalities can be highlighted.

Although this freshly presented metric can play a significant role in signalling the critical and risky states when combined with the action gap or the advising measure, it still maintains all the flaws of the static measures. A possible solution to adapt these static metrics to a dynamic version is stated in the subsequent section.

## 5.3.2   IQVA: The Iterated Q-Value Approach

While the previously examined metrics are relevant and useful in their specific contexts, none of them take into account the importance originated by the recurrent visitations of special states in several episodes. This characteristic is principally marking situations of bottlenecks and highly influential states that can majorly change the outcome of a solution. However, these states can not be recognized by the classical importance measures due to their estimations based only on the Q-values of that occasion. Moreover, with an approach completely built on the static state-action value function, a higher ranking is assigned to states that present a wide gap between the best action and another one (depending on the metric used). Nevertheless, it has not been considered whether the state is just rarely visited or a regularly encountered state that can affect the outcome of multiple trajectories. Hence, it could be beneficial to couple both visions in order to rescale the effect of the importance based on the relevance of the best Q-value and its frequency of appearance.

To this aim, an Iterated Q-Value Approach (IQVA) is presented in Algorithm 5.1. This iterated algorithm combines classical static metrics with the visitation frequencies to produce a measure able to satisfy both the conditions discussed above. Given the application of the standard importance measures, it is fundamental to assume that an initial Q-value function approximator is available. After this brief introduction, the complete algorithm is examined in depth.

At the beginning of the training or testing phase, a column vector $I \in \mathbb{R}^{|\mathbb{S}|}$, having the dimension equal to the number of states in the MDP tackled, is initialized. Then, for each episode until the maximum number $E \in \mathbb{N}$, the usual RL cycle is repeated. If the RL agent is still being trained, then the Q-value function approximator has to be first updated. Thereby, the Q-values can be used for the estimation of the importance to assign at the actual state instantiation $s_t = \bar{s} \in \mathbb{S}$. Every time that a state is visited, a contribution depending on a specific standard metric $I_M(\cdot)$ is added to its previous importance associated. Mathematically, the iterative updating rule is defined as follows:

$$I(\bar{s}) \leftarrow I(\bar{s}) + \hat{\gamma}^t I_M(\bar{s}) \tag{5.23}$$

where $\hat{\gamma} \in (0, 1]$ is the discount importance factor. This parameter, together with its exponent $t$ indicating the time step of the realization of the state $\bar{s}$, is utilized to decrease the influence of the importance measures for states that appear far in the future. Indeed, at the end of the episodes, the states visited by the RL agent are closer to the end goal; consequently, the Q-value differences achieve larger gaps. So, the adoption of this smoothing factor can balance the aforementioned issue. This applied idea comes from the rule-of-thumb technique of *exponential smoothing*: a method for assigning larger weights to more recent observations [80]. This updating procedure is iterated at each time step and every importance value for the visited states is increased. Finally, after completing the total number of episodes $E$, the importance vector obtained is normalized by dividing by $E$. In this way, the computed importance measures can be balanced with the number of episodes, eliminating the possibility of having exploding measurements. In fact, by dividing by the total amount of episodes, the remaining component is a combination of the probability of visiting that state and the relative smoothed Q-value-based metric.

---

**Algorithm 5.1** IQVA [158]

---

**Require:** Number of Episodes $E \in \mathbb{N}$, Discount importance factor $\hat{\gamma} \in (0, 1]$, Q-value function approximator.
1: Initialization of the importance measure

$$I(s) \leftarrow 0, \quad \forall s \in \mathbb{S}$$

2: **for** $episode = 1, \ldots, E$ **do**
3:     **for** $t = 0, 1, \ldots, T$ **do**
4:         Observe the state instantiation $s_t = \bar{s}$.
5:         **if** During training **then**
6:             Update Q-value function approximator
7:         **end if**
8:         Update Importance vector:

$$I(\bar{s}) \leftarrow I(\bar{s}) + \hat{\gamma}^t I_M(\bar{s}).$$

9:     **end for**
10: **end for**
11: Normalization:

$$I(s) \leftarrow \frac{I(s)}{E}, \quad \forall s \in \mathbb{S}.$$

12: **return** $I(s), \forall s \in \mathbb{S}$.

---

The noticeable qualities of this method do not rely only upon the ability to couple together the standard Q-value-based metrics with the visitations but also on the possibility of applying it in both the training and testing phase of an RL agent. Indeed, it can highlight the fundamental states based on the different moments of application. If used during the training, the spotted situations concern the highly visited states that have attained good results in the solution of the task. Furthermore, it is helpful to observe that the employment of this metric during the learning phase could be used to limit the exploration. While updating the different Q-values, a measure determining the ranking of the states is already available. So, the training could be extensively performed in those relevant positions to find possible better trajectories by exploiting the already collected information. On the other hand, the evaluation of the IQVA algorithm during the testing process can provide insights into the final perception of the agent. In this scenario, the important states detected are strictly related to the often visited states of optimal trajectories.

As it is shown in the following section, major differences between the two settings can be outlined in the computational experiments. Nevertheless, the consistency of the suggested approach is verified in the reported evaluation study.

### 5.3.3   Computational Experiments

The computational experiments carried out are performed in three grid-world environments shown in Fig. 5.1: Lava-lake, Key-door, and Taxi. Further information regarding these environments can be found respectively in Appendix A.4–A.5–A.1 in the Electronic Supplementary Material.

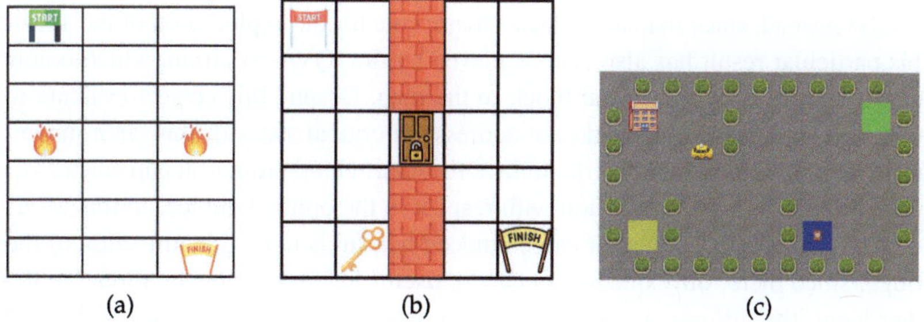

**Fig. 5.1** Environment used for the test of IQVA algorithms: (a) Lava-lake, (b) Key-door, and (c) Taxi

For each of these MDPs, a SARSA or Q-learning agent have been trained to accomplish the corresponding tasks. Moreover, the IQVA measurements were evaluated both during the learning process and after it. Hence, the reported evaluations will be indicated with the labels "train" and "post" for each of the occasions mentioned above. Additionally, the IQVA algorithm has to consider a static metric, meaning that it can transform the standard measures into visit-aware measurements. Thus, three different evaluations were associated with the IQVA values: IQVA-A when applying the advising importance, IQVA-AG for the utilization of the action gap, and IQVA-D in the case of using the difference metric. Then, the computed values are shown in heatmaps representing the different importance for each location. Precisely for this reason, the environments considered for the computational evaluation of the proposed methodologies are grid-worlds. Indeed, this simplifies the visualization and the comprehension of the results, helping in the recognition of the best measures for each problem. A total of 9 heatmaps, consisting of three kinds of qualities (standard, IQVA-train, and IQVA-post) and three types of metrics (Advising, Action Gap, and Difference), are discussed in the following.

Beginning with the Lava-lake problem, the RL algorithm trained for solving this task is SARSA using a total of 10000 episodes and $\gamma = 0.95$. For the evaluation of IQVA, the same amount of episodes and discount factors were used in both the "train"- and "post"-computations, i.e., $E = 10000$ and $\hat{\gamma} = \gamma = 0.95$.

IQVA-A facilitates discerning the most critical situations during learning. This becomes evident by the results shown in Fig. 5.2, when looking at the states near lava in the upper part of the map (that is the ones with values close to 0.05). On the other hand, the same is not attainable by the static advising measure since it strictly highlights the states that are close to the goal first and only after including these risky situations. In the "post"-evaluation phase, biggest importance is given to the bridge location instead, since that state plays a major role for the exploitation of the agent. This particular result has also been achieved by the IQVA-AG-train, which points out the relevance of the central block in the map. Despite this correct evaluation, the action gap measurements do not address the critical states of this environment due to the definition of the metric. In fact, the idea behind the action gap consists of evaluating the second-best action with respect to the optimal choice. In this MDP, the only circumstance where it can be indeed helpful is in the identification of the bridge, since there, only one action can be useful for finding the solution. On the other hand, the difference metric is able to detect the high-risk states in the first half of the map and the central bridge position in its standard form. Furthermore, it can give also importance to the two states adjacent to the lava in the second half of the environment. However, the importance achieved smaller absolute values. The IQCA-D results are similar to the IQVA-A ones. Hence, in this experiment, the static difference metric outperforms the iterated versions.

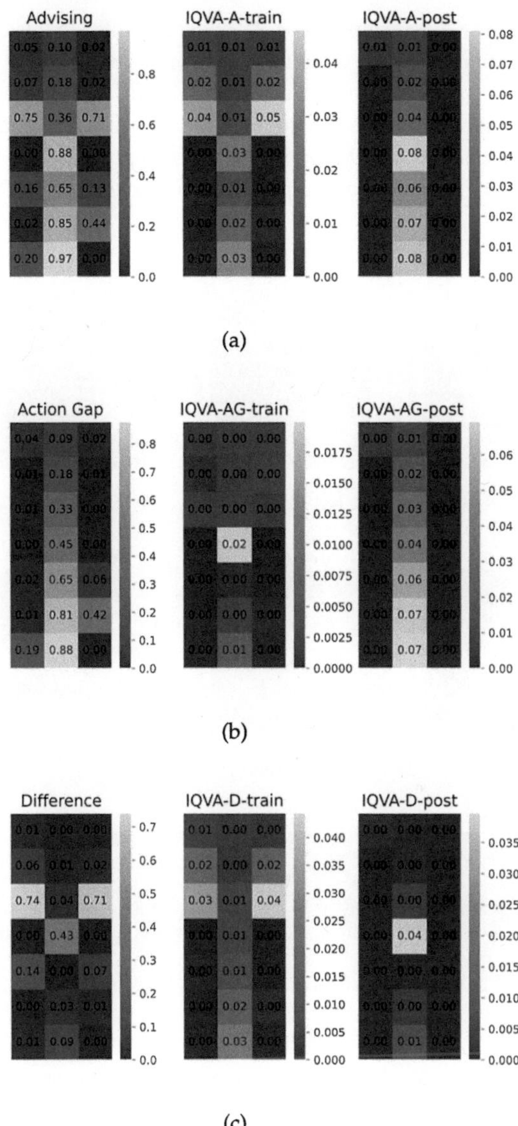



**Fig. 5.2** Heatmaps for the Lava-lake environment: measures relative to (a) Advising, (b) Action gap, and (c) Difference

Moving to the Key-door problem, the RL agent is trained in this context using a Q-learning method. In detail, the learning process, and consequently the test phase, lasted for $E = 1000$ episodes and used $\gamma = \hat{\gamma} = 0.95$. The obtained importance measures are visualized, for this environment, in two different situations: before picking up the key and after grabbing it. The calculated values are then shown in the usual heatmaps in Fig. 5.3 for the pre-pick-up scenario and in Fig. 5.4 for the post-pick-up one.

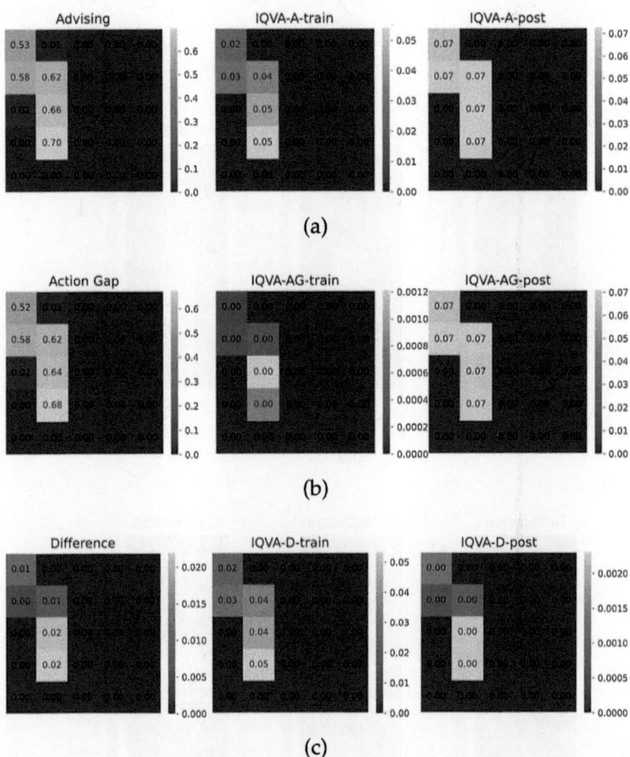

**Fig. 5.3** Heatmaps for the Key-door environment (pre-pick up key): measures relative to (a) Advising, (b) Action gap, and (c) Difference

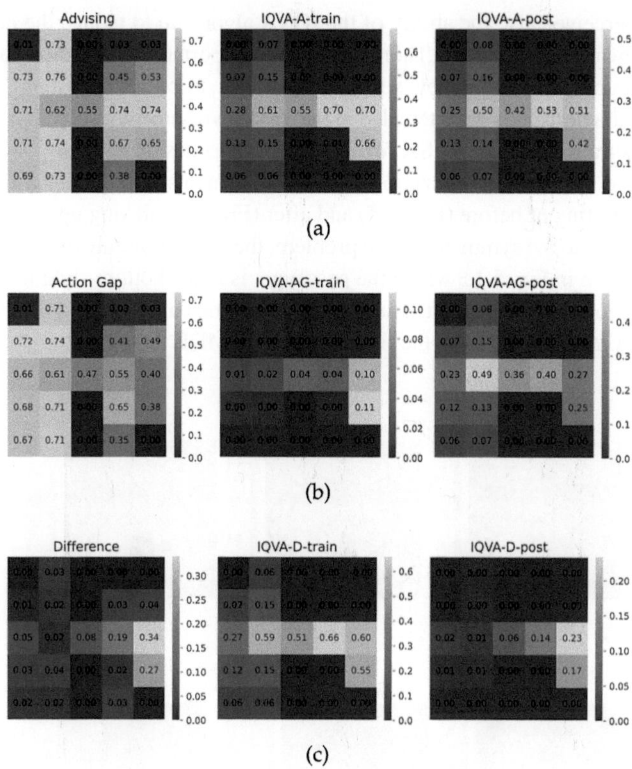

**Fig. 5.4** Heatmaps for the Key-door environment (post-pick up key): measures relative to (a) Advising, (b) Action gap, and (c) Difference

The analysis of the calculated importance measures for the instants before taking the key does not suggest any obvious identification of relevant states. The same can not be said for the second phase, which starts right after picking the key. In this frame, the metrics from the literature give more importance to the states that are in the first room of the map, instead of awarding a higher relevance to the door position or the goal state. The only static measure that can attain this requirement is the difference measure, which identifies the states close to the goal as the most fundamental ones. In general, the aforementioned issues are solved by the usage of the visitation count applied by the IQVA algorithm. The outputs from the IQVA method positively reward the central bottleneck of the problem. Hence, for this MDP it can be perceived also the ability of the IQVA algorithm to reduce the number of highly relevant states detected. Thereby, the set of important states is more expressive since it groups only the valuable situations that have to be borne in mind.

This characteristic is depicted also in the Taxi environment results. For this task, an agent is trained using SARSA for $E = 5000$ episodes using $\gamma = \hat{\gamma} = 0.95$. As done for the Key-door visualizations, the measurements are illustrated in the two frames consisting of before (Fig. 5.5) and after (Fig. 5.6) picking up the passenger. Moreover, due to the symmetry of the problem, the evaluations are only reported for the case shown in Fig. 5.1c, where the passenger is in the bottom-right location and the destination is in the top-left. Thus, the examination can start from the evaluation of the outcomes derived from the first stage of the MDP.

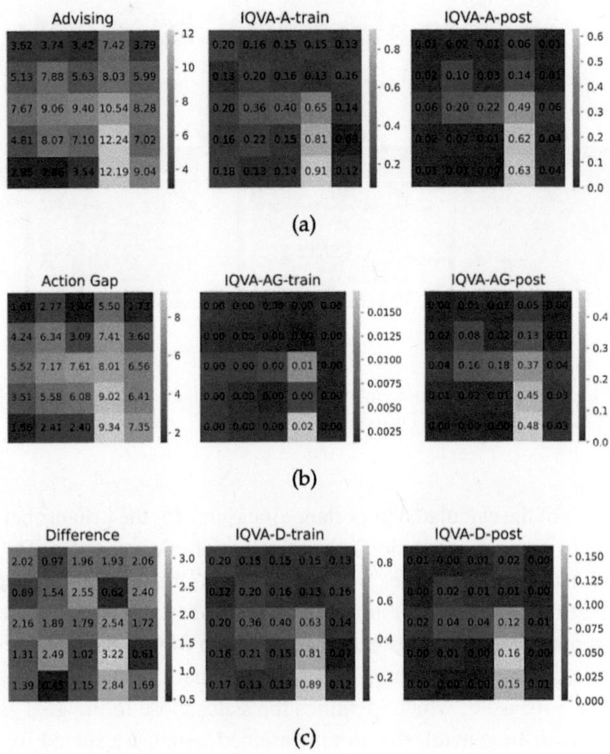

**Fig. 5.5** Heatmaps for the Taxi environment (pre-pick up): measures relative to (a) Advising, (b) Action gap, and (c) Difference

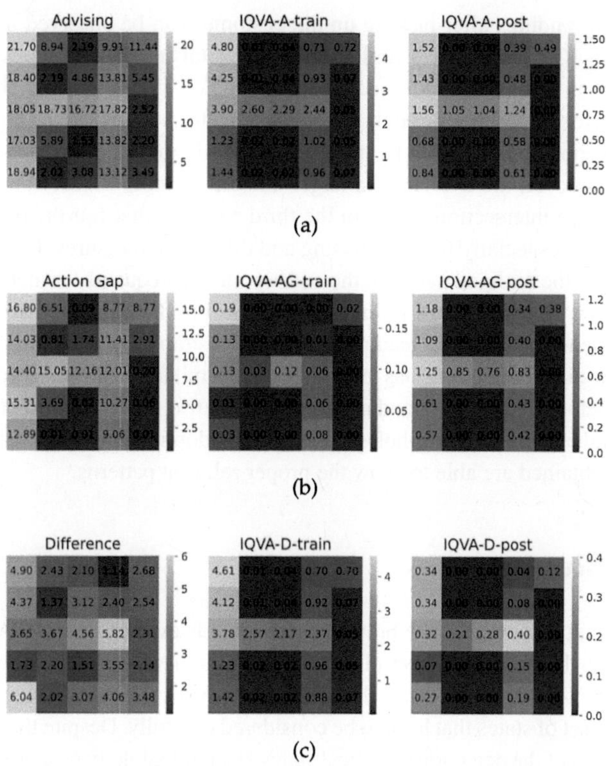

Fig. 5.6 Heatmaps for the Taxi environment (post-pick up): measures relative to (a) Advising, (b) Action gap, and (c) Difference

In Fig. 5.5, as already stated before, the IQVA methodology allows the user to visualize a smaller set of states recognized as important. By doing so, the validation can be done quickly and without any large loss of time. More in-depth, in all the combinations of measures and phases, the correct path, that has to be followed to pick up the passenger, can be discriminated. Another remark in this scenario concerns the seemingly random measurements of the difference metric. While for the advising and action gap values a rule-of-thumb idea can be determined for the recognition of the important states detected (the closer to the passenger the higher the relevance measured), it can not be stated the same for the difference metric since no apparent information can be deduced. However, the IQVA-D values are instead correctly addressing the sequence of states leading to the client's destination, thereby improving the efficiency of the difference measure too.

Interesting findings after picking up the customer can be observed in Fig. 5.6. By using the IQVA method, a limitation of the total number of states regarded as important is achieved. Additionally, for the IQVA measurements shown, the central trajectories passing through the environmental bottleneck are highlighted. In fact, these positions are used by a large amount of optimal policies to reach the end goal. Moreover, the overlap of a larger amount of these trajectories is clearly identified by the intersection points in the third row and first-fourth column. This facet is evident, especially for the advising and difference measures. For the action gap outcomes, the outlined path and the corresponding frequently visited states are well-addressed only in the post-training evaluation. For the training phase, only the optimal sequence of locations leading to the goal is emphasized. A final comment regarding the difference metric has to be discussed: similarly to the results calculated in the pre-pick-up phase, the standard difference metric seems to present difficult-to-understand outcomes. Nevertheless, through the adoption of the IQVA algorithm, the outputs obtained are able to show the proper relevant patterns.

## 5.3.4   Discussion

From the results reported in the previous section, it is evident how IQVA can precisely address the important states (Taxi), and not assign relevance to the less visited situations (Key-door). In this way, the user that analyzes the outcome can focus on a smaller subset of states that have to be considered carefully. Despite the satisfying results in terms of the detection of critical states, the methodology proposed presents the same flaws connected to the adoption of Q-value-based metrics. In particular, it is strictly dependent on the trained agent. Nonetheless, this aspect can be viewed positively since it can provide a specific interpretation for the trained RL agent.

On the same thread, the application of IQVA to multiple kinds of RL methodologies has shown no differences. In both the cases of off-policy (Q-learning) and on-policy (SARSA), the iterated algorithm achieved accurate estimations. Hence this approach can be recognized as independent of the type of technique used for the solution of the task. Indeed, no knowledge about the agent is relevant for the IQVA algorithm, except the availability of the Q-values.

For the experiments presented in this section, a multitude of metrics have been compared with each other. This particular choice is motivated by the ultimate goal of this chapter being the discovery of the distal states, meaning that it is preferable to recognize these specific states with only one kind of measurement. Despite the IQVA attaining the best evaluations in this task, the choice of a metric that could be applied in general is not trivial. For this reason, it is helpful to define a large set of metrics that have to be empirically tested in the problem that has to be completed.

Only after this evaluation stage, the optimal choice is visible. However, this is not a sporadic issue that appears in this exceptional scenario. In fact, the selection of the correct metric for the recognition of central features is a well-known problem in the networks analytics field [249].

Remaining in the same mentioned area, in the next section a graph-based approach is derived through the employment of advanced network structures.

## 5.4  Extending the Graph-based Metrics to Multigraphs and Multiplexes

The main issue regarding graph-based metrics proposed in the literature concerns the impossibility of easily including the information about the rewards obtained at each time step. In this way, it is complex to associate the centrality measures computed to a meaning connected to the quality of the solutions. This problem becomes more critical in a particular kind of environment: specific MDPs with a large variety of reward values. In fact, in the opposite cases of MDPs where the reward function is sparse, i.e., only in a few situations there is a reward different from 0, the application of the unweighted approaches reported in the previous overview can still efficiently represent the optimality of the policies. Moreover, the simple methodologies using directed unweighted graphs can already identify the relevant states for the problems where it is fundamental to complete the task in the minimum amount of time.

It must be noted that environments inspired by real-world applications used in RL contexts present a high complexity in terms of reward functions [67]. Thus, an approach that can take into account this specificity is required. In the subsequent sections, two methodologies are developed in order to include the reward aspects into the classical directed graph by the usage of more advanced network structures: the multigraphs and multiplex networks.

### 5.4.1  The Reconstructed Transition Multigraph and its Centrality Measures

To clearly state the definition of the novel structures adopted for the evaluation of the important states, it is essential to first describe some preliminary notations. Throughout this section, it is assumed the existence of a set of trajectories (states, actions, and rewards) $\mathcal{T} = \{\tau_p\}_{p=1,\dots,E}$ which will be used in the creation of the reconstructed transition graphs. From each of these trajectories $\tau_p = (s_{0,p}, a_{0,p}, r_{1,p}, \dots, s_{T_p-1,p}, a_{T_p-1,p}, r_{T_p,p}, s_{T_p,p})$, the set of all the states visited during the trajectory (without repetitions) is denoted by

$$\$(\tau_p) = \{s_{t,p} \in \tau_p \mid s_{t,p} \neq s_{t',p}, \ \forall t' < t\} \tag{5.24}$$

and the set of the ordered triplets state-action-next state (without repetitions) by

$$\$A\$(\tau_p) = \{(s_{t,p}, a_{t,p}, s_{t+1,p}) \mid (s_{t',p}, a_{t',p}, s_{t'+1,p}) \neq (s_{t,p}, a_{t,p}, s_{t+1,p}),$$
$$s_{t',p}, a_{t',p}, s_{t'+1,p}, s_{t,p}, a_{t,p}, s_{t+1,p} \in \tau_p, \ \forall t' < t\}. \tag{5.25}$$

These concepts can be extended to the complete set of trajectories as follows:

$$V^M(\mathcal{T}) = \bigcup_{p=1}^{E} \$(\tau_p), \tag{5.26}$$

$$E^M(\mathcal{T}) = \bigcup_{p=1}^{E} \$A\$(\tau_p), \tag{5.27}$$

where, trivially, there are no repetitions in the elements of the union. With these two sets, it is possible to discriminate directly between the states that have been visited and also the ordered chains of state-action-next state: They will play a fundamental role in the mathematical structure analyzed here: the *Reconstructed Transition Multigraph*. In detail, $V^M(\mathcal{T})$ denotes the set of the visited states represented as nodes (or vertices), and $E^M(\mathcal{T})$ denotes the set of the performed actions indicated as parallel edges.

From the equations above, it can be noticed that the vertices are indicated as the single states visited during the training or test phase of an RL agent, while triplets are considered for the edges and the calculation of the corresponding weights. Nevertheless, this notation is heavy and not easy to grasp at first glance. Hence a simplification for the practical observation that will be derived is proposed.

For the vertices, the time step at which the state has been evaluated is dropped and instead a different counting index is introduced. This allows identifying diverse kinds of instantiations of the states. For example, the first state that is visited is denoted as $s_1 \in V^M(\mathcal{T})$, then, the next different state visited will be represented as $s_2$ and so on.

A more compact notation is also used for the edges, which have been initially represented as triplets, but now are denoted by $a_{i,j}^k \in E^M(\mathcal{T})$ where $a_{i,j}^k = (s_i, a_k, s_j)$ indicates the transition from the state instantiation $s_i$ to the next state instantiation $s_j$ using the action $a_k$. In this case, the index of the action is not relative to the time step but to the specific kind of action done. Thus, all the parallel edges between $s_i$ and $s_j$ can be obtained varying $k = 1, \ldots, |\mathbb{A}|$.

Lastly, associated with each edge $a_{i,j}^{k}$ is the weight $w_{i,j}^{k}$ that is defined as the output of the function $w$ in composition with the reward function $r$ using the edge information, i.e., $w(r(a_{i,j}^{k})) = w(r_{i,j}^{k})$ where $r_{i,j}^{k}$ is the reward observed when passing from state $s_i$ to state $s_j$ by acting $a_k$. In this way, the last component of the Reconstructed Transition Multigraph can be introduced:

$$W_w^M(\mathcal{T}) = \left( \left( W_w^M(\mathcal{T}) \right)_{i,j}^{k} \right)_{i,j=1,...,|V^M(\mathcal{T})|}^{k=1,...,|\mathbb{A}|} \tag{5.28}$$

with

$$\left( W_w^M(\mathcal{T}) \right)_{i,j}^{k} = \begin{cases} w(r(a_{i,j}^{k}) & \text{if } a_{i,j}^{k} \in E^M(\mathcal{T}), \\ 0 & \text{otherwise,} \end{cases} \tag{5.29}$$

where $W_w^M(\mathcal{T}) \in \mathbb{R}^{|V^M(\mathcal{T})| \times |V^M(\mathcal{T})| \times |\mathbb{A}|}$ is the tensor of the reward-transformed weights defined by the real function $w : \mathbb{R} \to \mathbb{R}$ and $\left( W_w^M(\mathcal{T}) \right)_{i,j}^{k} \in \mathbb{R}$ indicates its component of indexes $i$, $j$, $k$. Here, it is assumed that if there are no edges between two nodes, then the corresponding weight is 0.

With this remaining piece of information, all the preliminary notions have been described. Hence, the mathematical definition of the Reconstructed Transition Multigraph is presented.

**Definition 5.2**   *Given a set of trajectories* $\mathcal{T} = \{\tau_p\}_{p=1,...,E}$ *performed in a MDP* $(\mathbb{S}, \mathbb{A}, r, \mathbb{T}, \gamma)$, *the Reconstructed Transition Multigraph* $G_w^M(\mathcal{T})$ *is defined as follows:*

$$G_w^M(\mathcal{T}) = (V^M(\mathcal{T}), E^M(\mathcal{T}), W_w^M(\mathcal{T})). \tag{5.30}$$

*Here* $V^M(\mathcal{T})$ *is as in Equation* (5.26) *and denotes the set of the visited states represented as nodes (or vertices),* $E^M(\mathcal{T})$ *is as in Equation* (5.27) *and denotes the set of the performed actions indicated as parallel edges, and* $W_w^M(\mathcal{T})$ *is as in Equation* (5.28) *and denotes the set of the rewards-transformed weights defined by the real function* $w : \mathbb{R} \to \mathbb{R}$.

This particular formulation has been adapted from the classical concept of reconstructed transition graphs to the case of the multigraphs. In fact, in the Reconstructed Transition Multigraph case, an edge is defined not only by its starting and ending node (state) but by also the action that has been performed. Thus, the specific action selected indicates the label of the edge.

Other than the structural enhancements obtained by the possibility of representing the self-loops and the parallel edges corresponding to different actions, the principal advancement lies in the inclusion of the rewards in the graph. However, due to the complexity of well-defining the centrality measures in weighted multigraphs, we employed a specific mapping function for the rewards in order to transform them into correctly shaped weights for each edge. A specific analysis of several scenarios for these reward mappings is performed in the following pages.

Before analyzing the choices for the weight functions, it is necessary to describe the procedure of creation of the Reconstructed Transition Multigraphs. To define the Reconstructed Transition Multigraph, two approaches can be utilized, depending on the time of creation: when the Reconstructed Transition Multigraph is built while the RL agent is acting in the environment or by directly using a set of trajectories stored. For the former scenario, Algorithm 5.2 can be employed.

---

**Algorithm 5.2** Creation of Reconstructed Transition Multigraph during RL acting [160]

---

**Require:** Number of Episodes $E \in \mathbb{N}$, Policy $\pi$.
1: Initialize empty graph $G_w^M = (V^M, E^M, W_w^M) = (\emptyset, \emptyset, \mathbf{0})$
2: **for** $episode = 1, \ldots, E$ **do**
3:      Initialize state $s_i$
4:      **if** $s_i \notin V^M$ **then**
5:          $V^M \leftarrow V^M \cup \{s_i\}$                    $\triangleright$ Add vertex $s_i$
6:      **end if**
7:      **for** $t = 0, 1, \ldots, T$ **do**
8:          Take action $a_k = \pi(s_i)$ and observe the next state $s_j$ and reward $r_{i,j}^k$
9:          **if** $s_j \notin V^M$ **then**
10:          $V^M \leftarrow V^M \cup \{s_j\}$               $\triangleright$ Add vertex $s_j$
11:          **end if**
12:          **if** $a_{i,j}^k = (s_i, a_k, s_j) \notin E^M$ **then**
13:          $E^M \leftarrow E^M \cup \{a_{i,j}^k\}$       $\triangleright$ Add edge from $s_i$ to $s_j$ with label $k$
14:          Set the associated weight $w(r_{i,j}^k) = w(r(s_i, a_k, s_j))$
15:          $(W_w^M)_{i,j}^k \leftarrow w(r_{i,j}^k)$            $\triangleright$ Update weight $w(r_{i,j}^k)$
16:          **end if**
17:          Update state $s_i \leftarrow s_j$
18:      **end for**
19: **end for**
20: **return** $G_w^M$.

It can be beneficial to discuss a simple example of the latter case. In general, the final approach that is explained now can be applied in both scenarios, since the trajectories can be memorized and, then, used for constructing the Reconstructed Transition Graph.

Given the following set of trajectories:

$$\mathcal{T}_e = \{\tau_1 = (s_1, a_1, r^1_{1,2}, s_2, a_2, r^2_{2,3}, s_3, a_1, r^1_{3,4}, s_4, a_2, r^2_{4,3}, s_3),$$
$$\tau_2 = (s_1, a_3, r^3_{1,2}, s_2, a_3, r^3_{2,2}, s_2, a_1, r^1_{2,4}, s_4, a_2, r^2_{4,3}, s_3)\} \tag{5.31}$$

it is possible to generate the elements of the Reconstructed Transition Graph according to the subsequent procedure. First, the set of the visited states is produced by indicating all the state values found in each trajectory: $\mathbb{S}(\tau_1) = \{s_1, s_2, s_3, s_4\}$, $\mathbb{S}(\tau_2) = \{s_1, s_2, s_4, s_3\}$. Hence, $V^M(\mathcal{T}_e) = \{s_1, s_2, s_3, s_4\}$ is obtained. Then, the set of ordered triplets $\mathbb{SAS}$ is constructed by looking at the state-action-next state chain of information in the trajectories. Specifically to this example, it is obtained:

$$\mathbb{SAS}(\tau_1) = \{(s_1, a_1, s_2), (s_2, a_2, s_3), (s_3, a_1, s_4), (s_4, a_2, s_3)\}, \tag{5.32}$$
$$\mathbb{SAS}(\tau_2) = \{(s_1, a_3, s_2), (s_2, a_3, s_2), (s_2, a_1, s_4), (s_4, a_2, s_3)\}, \tag{5.33}$$

thus, the set of parallel edges $E^M(\mathcal{T}_e)$ is

$$E^M(\mathcal{T}_e) = \{(s_1, a_1, s_2), (s_2, a_2, s_3), (s_3, a_1, s_4), (s_4, a_2, s_3),$$
$$(s_1, a_3, s_2), (s_2, a_3, s_2), (s_2, a_1, s_4)\} \tag{5.34}$$
$$= \{a^1_{1,2}, a^2_{2,3}, a^1_{3,4}, a^2_{4,3}, a^3_{1,2}, a^3_{2,2}, a^1_{2,4}\} \tag{5.35}$$

where the edges have been reported in both the extended and compact forms. It must be noticed that the set of ordered triplets for the trajectories $\tau_1$ and $\tau_2$ had 4 elements each, while the edge set has a cardinality of 7. This reduction is due to the presence of a shared ordered triplet between the two trajectories, which is not repeated in the union. Finally, the weights associated with these edges are computed using a function $w$ and the rewards earned for each triplet of state-action-next state. So, the following tensor of the weights is obtained:

$$\left(W^M_w(\mathcal{T})\right)^1_{1,2} = w(r^1_{1,2}), \quad \left(W^M_w(\mathcal{T})\right)^1_{1,2} = w(r^2_{2,3}), \quad \left(W^M_w(\mathcal{T})\right)^1_{1,2} = w(r^1_{3,4}),$$
$$\left(W^M_w(\mathcal{T})\right)^1_{1,2} = w(r^2_{4,3}), \quad \left(W^M_w(\mathcal{T})\right)^1_{1,2} = w(r^3_{1,2}), \quad \left(W^M_w(\mathcal{T})\right)^1_{1,2} = w(r^3_{2,2}),$$
$$\left(W^M_w(\mathcal{T})\right)^1_{1,2} = w(r^1_{2,4}) \tag{5.36}$$

and equal to 0 for the remaining components of $W^M_w(\mathcal{T})$.

With this information, the graph can be generated by constructing the vertices using $V^M(\mathcal{T})$ and linking them together following the edges listed in $E^M(\mathcal{T})$. The final result is the Reconstructed Transition Multigraph, shown in Fig. 5.7.

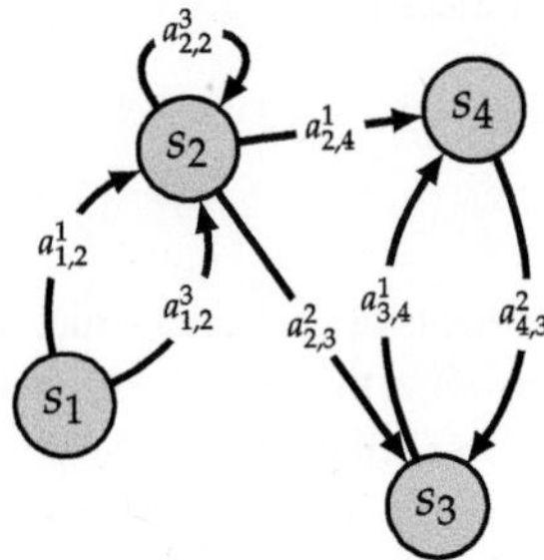


**Fig. 5.7** Example of Reconstructed Transition Multigraph

A major aspect that has been left behind is the description of the choice of the function $w$. The correct formulation of the weights depends on the selection of $w$ and, consequently, the appropriate solutions determined by the centrality measures computed. However, this is not a trivial choice since multiple assumptions are required to be satisfied in order to maintain the well-definiteness of the centrality metrics. Therefore, in the next pages, the focus is shifted to finding the correct mappings for the weights, in order to have well-defined three classical centrality measures: betweenness, closeness, and eigenvector centrality.

For the first measure, it is fundamental to transform the rewards into distances. While it is possible to have negative distances, the only situation that has to be avoided corresponds to the presence of negative cycles. Hence, the simplest decision is to consider $w(r) = -r$. Indeed, this simple transformation leads to an important new result.

**Theorem 5.4** *Given a Reconstructed Transition Multigraph $G_w^M(\mathcal{T})$, created from the set of trajectories $\mathcal{T}$ extracted from a MDP, if $w(r) = -r$ and $\gamma = 1$, then solving the shortest path problem in $G_w^M(\mathcal{T})$ is equivalent to finding the optimal trajectory in the MDP.*

**Proof.** Let $s, g \in V^M(\mathcal{T})$ be two vertices in the Reconstructed Transition Multigraph and the associated shortest path $\bar{x}(s, g) = (\bar{x}_{i,j}^k(s, g))_{(i,j,k)\in\mathcal{I}_E}$, where $\mathcal{I}_E$ denotes the set of indexes of $E^M(\mathcal{T})$. In detail, each binary variable $x_{i,j}^k(s, g)$

represents the choice of edge $a_{i,j}^k$ in the shortest path connecting $s$ to $g$ when $x_{i,j}^k(s,g) = 1$; otherwise $x_{i,j}^k(s,g) = 0$ when edge $a_{i,j}^k$ has not been selected. Hence, from the length of the shortest path changed in sign, the following chain of equations can be computed:

$$-\sum_{(i,j,k)\in\mathcal{I}_E} \left(W_w^M(\mathcal{T})\right)_{i,j}^k \bar{x}_{i,j}^k(s,g) = -\min_{x(s,g)\in\{0,1\}^{|\mathcal{I}_E|}} \sum_{(i,j,k)\in\mathcal{I}_E} w(r_{i,j}^k)x_{i,j}^k(s,g) \tag{5.37}$$

$$= -\min_{\tau(s,g)} \sum_{(i,j,k)\in\mathcal{I}_{\tau(s,g)}} w(r_{i,j}^k) \tag{5.38}$$

$$= -\min_{\tau(s,g)} \sum_{(i,j,k)\in\mathcal{I}_{\tau(s,g)}} -r(a_{i,j}^k) \tag{5.39}$$

$$= -\min_{\tau(s,g)} -\sum_{t=0}^{T_{\tau(s,g)}} r_t \tag{5.40}$$

$$= \max_{\tau(s,g)} \sum_{t=0}^{T_{\tau(s,g)}} r_t \tag{5.41}$$

where Equation (5.37) is the definition of the shortest path value. Equation (5.38) is verified when only the trajectories going from $s$ to $g$ are considered; hence, $x_{i,j}^k(s,g) = 1, \forall(i,j,k) \in \mathcal{I}_{\tau(s,g)}$, where $\mathcal{I}_{\tau(s,g)}$ is the set of indices of states and actions in the trajectory $\tau(s,g)$. Equation (5.39) follows from $w(r) = -r$ and Equation (5.40) is derived by updating the notation following that of the MDPs and by defining $T_{\tau(s,g)}$ as the terminal time step in the trajectory $\tau(s,g)$. Subsequently, Equation (5.41) holds by transforming the min into a max through the cancellation of the minus signs. Since the return is calculated with $\gamma = 1$, the maximum of the objective function of the MDP has been found. Moreover, this is verified for all starting states $s$ and the goal states $g$, due to their general aspects.

The last remark concerns the identification of a bijection between the paths in the shortest path problem and trajectories in the MDPs. However, this property can be directly obtained from the definition of a Reconstructed Transition Multigraph (Definition 5.2). $\qquad\square$

With this result, the concept of the shortest path in Reconstructed Transition Multigraphs and optimal trajectory in MDPs can be coupled together. By doing so, it is straightforward to indicate the important states as the ones that have a high number of shortest paths passing through them. Indeed, these are the states that have numerous (near) optimal trajectories passing through them. Hence, the betweenness centrality has a well-defined meaning if applied in this particular context. Furthermore, it is interesting to notice that the presence of negative cycles in the Reconstructed Transition Graph is linked to the existence of positive reward cycles that can be exploited in MDPs. Thus, what appeared at the beginning to be a problem can now be a helpful source to detect possible inaccuracies in the definition of the MDPs.

The simplistic choice of the mapping for the weights provides satisfying outcomes for the betweenness. However, the same can not be stated for the other two centrality measures. The main issue that invalidates the application of this function for the closeness and eigenvector metrics is the presence of negative values derived from the positive rewards gained by the RL agent.

The first idea that comes to mind is to translate all the weights by subtracting the minimum. Through this transformation, all the resulting weights become non-negative. However, the resulting shortest path distance is completely distorted, inducing in this way a different interpretation of the importance that is no longer connected to the MDP. A simple example demonstrates this problem.

Consider the simple graph visualized in Fig. 5.8a. In this case, the shortest path between node $A$ and node $C$ consists of passing through $B$. The length of this path is 0, while, the edge connecting $A$ to $C$ has a weight of 0.5. However, this situation is completely different if we increase by 1 all the weights in order to have non-negative values. The resulting graph after the translation of the weights is shown in Fig. 5.8b. Here, it can be noted how the shortest path from $A$ to $C$ consists of using the edge $A \rightarrow B$. Indeed, the cost is only 1.5 while the other path has a total length of 2. Thus, the shortest path changed because of this simple increment of 1 unit for each weight.

A modification that does not lead to a change in the context of the shortest path is to map all the positive rewards to a small constant $\epsilon_C > 0$ s.t. $\epsilon_C \leq \min_{r>0, r \in \mathcal{T}} |r|$. Hence, the obtained weight function considered for the closeness centrality is defined as follows:

$$w_C(r) = \begin{cases} -r, & \text{if } r < 0 \\ \epsilon_C, & \text{if } r \geq 0 \end{cases}. \tag{5.42}$$

With this choice for the reward-based weights, the shortest path found in the multigraph is not extensively modified. In the previous example, it can be fixed $\epsilon_C = 0.4$ since the smallest reward is 0.5. Hence, the resulting graph maintains the same shortest path of the initial graph.

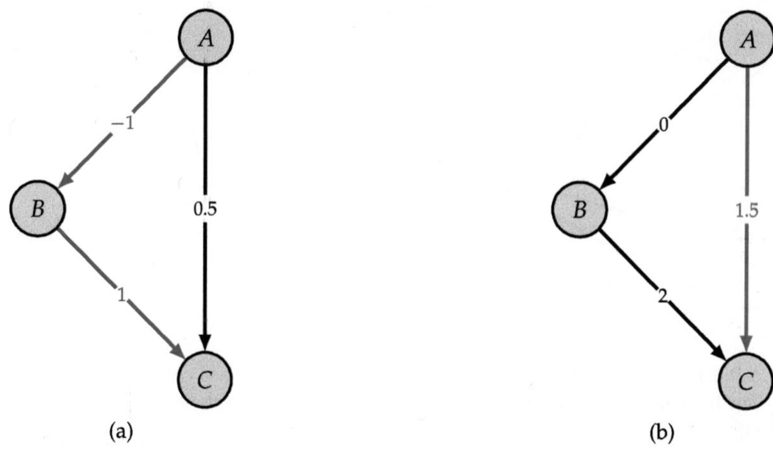



**Fig. 5.8** Example of how, by translating all the weights of a constant value, the shortest paths (highlighted) are modified: (a) graph before the translation, (b) graph after translating the weights of 1

Nevertheless, the condition of not being in a sparse-reward environment must hold. In fact, in those cases, the rewards equal to 0 are considered positive, thus, they are transformed in the constant value $\epsilon_C$. To avoid this issue, the application of these methodologies will be restricted to non-sparse-reward environments. Indeed, as already depicted in the literature review, multiple approaches for these situations can obtain a decent estimation of the important states, such as [171, 195, 223].

Additionally, another condition has to be ensured in order to maintain the same idea of shortest path in the Reconstructed Transition Multigraphs. This requirement concerns the particular structure of the MDPs addressed: positive rewards must be obtained in specific bottleneck edges that are the only connections to particular sub-goal nodes. A representation of this scenario is depicted in Fig. 5.9, where, in order to reach the sub-goal node $s_g$ it is necessary to pass in $s_{g-1}$ and take the action $a^k_{g-1,g}$. This setting is a common scenario in MDPs since the higher rewards are usually linked to the completion of a (sub-)goal. Moreover, the goal states are achieved through bottleneck actions, indicating that only an edge can be crossed to reach the terminal state.

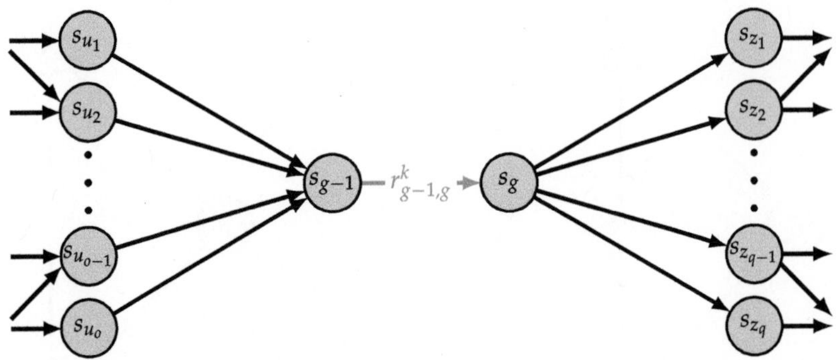

**Fig. 5.9** Example of the structural bottleneck edges condition used considered in Theorem 5.5. The bottleneck edge having $r^k_{g-1,g} > 0$ and connecting $s_{g-1}$ to the sub-goal $s_g$ is highlighted

An example of this phenomenon has already been encountered in the Taxi environment, where, to complete the task, the drop off action has to be performed at the correct location. The same concept is applied also to the sub-goal of picking up the passenger, which constitutes another example of the aforementioned structural circumstances; although the reward associated is still negative. Therefore, it must be remembered that sub-goals are fundamental states that lead to the positive completion of the overall problem, but they are not always characterized by positive incoming or outgoing edge weights.

Under these structural hypotheses, the following novel result can be proved, showing that truncating only these positive weights related to bottlenecks edges does not significantly influence the shortest paths in the Reconstructed Transition Multigraphs.

**Theorem 5.5** *Given the Reconstructed Transition Multigraphs $G^M_w(\mathcal{T})$, and $G^M_{wC}(\mathcal{T})$, the shortest paths between two states $s_0$ and $s_f$ are the same in both the multigraphs if for each positive reward $r^k_{g-1,g} > 0$ the associated edge $a^k_{g-1,g} \in E^M(\mathcal{T})$ is the unique in-coming edge of the (sub-)goal state $s_g$.*

**Proof.** First, it is proved that any shortest path from $s_0$ and $s_f$ (which is not necessarily a sub-goal) in $G^M_w(\mathcal{T})$ is a shortest path in $G^M_{wC}(\mathcal{T})$ too. Then, due to the arbitrary choice of $s_0$ and $s_f$, the proof holds for each couple of vertices. Furthermore, these arguments can be repeated for the opposite implication.

Let $P(s_0, s_f) = (s_0, s_{f_1}, \ldots, s_{f_o}, s_f)$ be a shortest path in $G^M_w(\mathcal{T})$ connecting state $s_0$ to state $s_f$. It is required to prove that $P(s_0, s_f)$ is also the shortest path between $s_0$ and $s_f$ in $G^M_{wC}(\mathcal{T})$. If there are no positive rewards along the shortest path, then the equality is trivially proved.

In the other circumstance, it is helpful to limit the discussion to the first chain of the shortest path, i.e., from the initial node until the first sub-goal. In fact, given the identification of the sub-goal to the bottlenecks for the trajectories in MDPs, all the paths from $s_0$ to $s_f$ have to pass through them. Moreover, these nodes have the function of connecting multiple parts of the graph, leading to being the only choices for moving from one component of the network to the other. Hence, the search for the shortest path between two nodes $s_0$ and $s_f$ in diverse components can be decoupled in the recognition of the shortest path between the starting node $s_0$ and the corresponding bottleneck vertex $s_g$, and the discovery of the shortest path between $s_g$ and $s_f$. This idea can be reiterated multiple times until the divided paths present only one sub-goal state with an incoming positive reward.

So, from now on, the path $P(s_0, s_g)$ is analyzed, where it is supposed that the only sub-goal state with incoming positive reward present is the final node $s_g$. Let $P_C(s_0, s_g) = (s_0, s_{j_1}, \ldots, s_{j_m}, s_{g-1}, s_g)$ be the shortest path in $G^M_{wC}(\mathcal{T})$. The goal consists of showing that the following equality holds:

$$l_{wC}(P_C(s_0, s_g)) = l_{wC}(P(s_0, s_g)) \tag{5.43}$$

where $l_g(P) = \sum_{(i,j,k)\in\mathcal{I}_P} g(r^k_{i,j})$ represents the length of path $P$ using the weight function $g$, under the set of indexes $\mathcal{I}_P$ of the path $P$.

From to the definition of the shortest path in $G^M_{wC}(\mathcal{T})$, it trivially follows:

$$l_{wC}(P_C(s_0, s_g)) \leq l_{wC}(P(s_0, s_g)), \tag{5.44}$$

since this inequality holds for all the paths connecting $s_0$ and $s_g$.

In order to complete the proof, it is required to show that

$$l_{wC}(P(s_0, s_g)) \leq l_{wC}(P_C(s_0, s_g)). \tag{5.45}$$

To accomplish this task, a proof by contradiction is performed. Suppose:

$$l_{wC}(P_C(s_0, s_g)) < l_{wC}(P(s_0, s_g)). \tag{5.46}$$

Obviously, since $w_C(r) = w(r)$, $\forall r < 0$, there exists at least an edge with a positive reward, otherwise, this strict inequality does not hold. In this case, the edge having a positive reward is only one from the preliminary description of the setting: $a^k_{g-1,g}$

(for a specific $k$ set in the hypothesis). So, Equation (5.46) can be algebraically explicated as:

$$\sum_{(i,j,k)\in \mathcal{I}_{P_C(s_0,s_g)}} w_C(r^k_{i,j}) < \sum_{(i,j,k)\in \mathcal{I}_{P(s_0,s_g)}} w_C(r^k_{i,j}). \qquad (5.47)$$

Now, the two paths are decomposed into a first sub-path between $s_0$ and the vertex $s_{g-1}$ connected to the ending node $s_g$ and the remaining edge linking $s_{g-1}$ to $s_g$. In this way, the latter weights can be replaced by the constant $\epsilon_C$ using the definition of $w_C$, since the associated reward is positive for the hypothesis. Hence, it holds:

$$\sum_{(i,j,k)\in \mathcal{I}_{P_C(s_0,s_{g-1})}} w_C(r^k_{i,j}) + \epsilon_C < \sum_{(i,j,k)\in \mathcal{I}_{P(s_0,s_{g-1})}} w_C(r^k_{i,j}) + \epsilon_C \qquad (5.48)$$

where $w_C$ can be also substituted with $w$ because all the remaining edges are associated with negative rewards. So,

$$\sum_{(i,j,k)\in \mathcal{I}_{P_C(s_0,s_{g-1})}} w(r^k_{i,j}) < \sum_{(i,j,k)\in \mathcal{I}_{P(s_0,s_{g-1})}} w(r^k_{i,j}) \qquad (5.49)$$

which represents the comparison between the lengths of the two paths:

$$l_w(P_C(s_0, s_{g-1})) < l_w(P(s_0, s_{g-1})). \qquad (5.50)$$

The above inequality leads to a contradiction since $P(s_0, s_{g-1})$ is part of a shortest path $P(s_0, s_g)$, so, it is also a shortest path in $G^M_w(\mathcal{T})$. Indeed, if this statement is true, then it is possible to replace $P(s_0, s_{g-1})$ in $P(s_0, s_g)$ with $P_C(s_0, s_{g-1})$ and obtain a shorter path; leading to the contradiction. Thereby, Equation (5.46) is contradicted, leading to the verification of Equation (5.45).

By repeating the same argument in the opposite direction, the theorem is proved.  $\square$

The previous theorem can be applied only under the assumption that there is only one specific vertex connected by an edge to the goal node. Despite this strong assumption, this result can be extended to other kinds of environments. In fact, as depicted in Fig. 5.10, if multiple edges connecting any node to the (sub-)goal state share the reward associated, then it is possible to show that the shortest paths are preserved. This statement can in this way enlarge the number of environments where the closeness centrality can be applied. An example of a problem included in this second class is the Key-door MDP. In this circumstance, it is indeed allowed to reach the (sub-)goal from multiple adjacency states. By doing so, the rewards obtained are always the same.

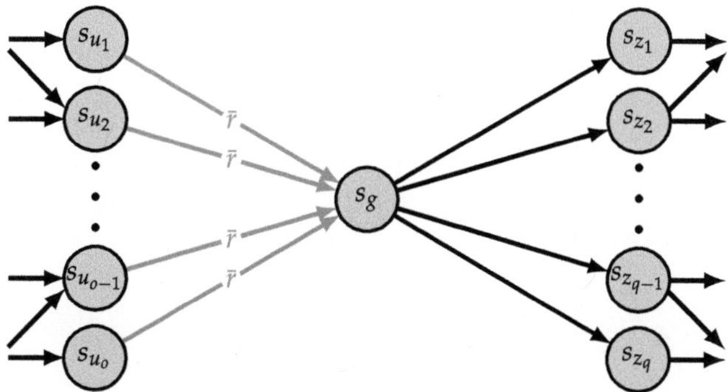

**Fig. 5.10** Example of the structural bottleneck node condition used considered in Theorem 5.6. The edges having positive equal reward $\bar{r} > 0$ and incoming to the bottleneck node $s_g$ are highlighted

After discussing these preliminary implications, the principal result is now outlined.

**Theorem 5.6** *Given the Reconstructed Transition Multigraphs $G_w^M(\mathcal{T})$, and $G_{wC}^M(\mathcal{T})$, the shortest paths between two states $s_0$ and $s_f$ are the same in both the multigraphs if for each positive reward $r_{i,g}^k > 0$ associated with the edge $a_{i,g}^k \in E^M(\mathcal{T})$ denoting an in-coming edge of the (sub-)goal state $s_g$, it holds that $r_{i,g}^k = r_{j,g}^h$, $\forall i, j, k, h,$ s.t. $a_{i,g}^k, a_{j,g}^h \in E^M(\mathcal{T})$.*

**Proof.** The proof is based on the same schematic idea considered in the proof of Theorem 5.5.

First, it is proved that any shortest path in $G_w^M(\mathcal{T})$ is a shortest path in $G_{wC}^M(\mathcal{T})$ too. Then, the opposite implication is proved using the same arguments. Using the same reasoning as in the previous proof, it is possible to focus the attention on the sub-paths leading to the first sub-goal. Let $P(s_0, s_g) = (s_0, s_{i_1}, \ldots, s_{i_l}, s_g)$ be the shortest path in $G_w^M(\mathcal{T})$ connecting vertex $s_0$ to $s_g$ and $P_C(s_0, s_g) = (s_0, s_{j_1}, \ldots, s_{j_m}, s_g)$ be the shortest path in $G_{w_C}^M(\mathcal{T})$ between the same nodes. It is required to show that

$$l_{w_C}(P(s_0, s_g)) \leq l_{w_C}(P_C(s_0, s_g)) \tag{5.51}$$

which can be done by contradiction. Suppose by contradiction that

$$l_{w_C}(P_C(s_0, s_g)) < l_{w_C}(P(s_0, s_g)), \tag{5.52}$$

where, also in this case, all the rewards earned following both the paths are negative other than the last one obtained passing from the second to last node to the (sub-)goal vertex, where $r_{i_l,g}^h = r_{j_m,g}^k > 0$ (for hypothesis). Hence, Equation (5.52) can be explicated as follows:

$$\sum_{(i,j,k)\in\mathcal{I}_{P_C(s_0,s_{j_m})}} w_C(r_{i,j}^k) < \sum_{(i,j,k)\in\mathcal{I}_{P(s_0,s_{i_l})}} w_C(r_{i,j}^k) \tag{5.53}$$

where the weight associated with the edge going to the sub-goal $s_g$ has been cancelled since it is equal to $\epsilon_C$ for both paths. So, now the sub-paths analyzed are $P_C(s_0, s_{j_m})$ and $P(s_0, s_{i_l})$. Then, bearing in mind that $r_{i_l,g}^h = r_{j_m,g}^k$ are the only positive rewards, the inequality can be updated as follows:

$$\sum_{(i,j,k)\in\mathcal{I}_{P_C(s_0,s_{j_m})}} w(r_{i,j}^k) - r_{j_m,g}^k < \sum_{(i,j,k)\in\mathcal{I}_{P(s_0,s_{i_l})}} w(r_{i,j}^k) - r_{i_l,g}^h \tag{5.54}$$

where the function $w_C$ has been replaced by $w$ since in the paths $P_C(s_0, s_{j_m})$, $P(s_0, s_{i_l})$ there are no other positive rewards, and the rewards are subtracted to both the sides without any problem due to their equivalence. In this way, it can be seen that both sides can be rewritten using the length function in $G_w^M(\mathcal{T})$ as:

$$l_w(P_C(s_0, s_g)) < l_w(P(s_0, s_g)), \tag{5.55}$$

which creates a contradiction since $P(s_0, s_g)$ is the shortest path in $G_w^M(\mathcal{T})$. Therefore, Equation (5.51) is verified.

Repeating this argument for the opposite implication leads to the completion of the proof. $\qquad\square$

So far, the requirements needed for the existence and well-definiteness of the closeness centrality metric consist of the positivity of the weights and the presence of particular structures in the reconstructed graphs. Compared to the assumptions required for the betweenness centrality, the hypotheses used here are stronger and more limiting. Moreover, even more stringent conditions have to be assumed for the eigenvector centrality in order to adopt it in the context of the Reconstructed Transition Multigraphs.

The first prerequisite is the non-negativity of the weights. To this end, it is possible to utilize the usual weighting function $w_C$. However, in this scenario, the weights indicate the relevance of the connection, so, higher values determine a major importance. Thus, the reciprocal $\frac{1}{w_C(\cdot)}$ must be utilized. With this measure, the evaluations obtained by the eigenvector centrality have the correct meaning. Nevertheless, it is still not sufficient to ensure the existence of this particular measure. In fact, in order to have a well-posed metric, the strong connectivity of the Reconstructed Transition Multigraph must be hypothesized. This specific condition is not in general satisfied for the proposed graphs. Not even the connectivity of the graph can be considered since there are multiple examples of MDPs generating a non-connected Reconstructed Transition Multigraph, e.g., the Taxi problem. The main problem in this situation concerns the directions of the edges: if the edges in the Reconstructed Transition Multigraph are undirected, then, by restricting the analysis only to one connected component at a time, it is possible to verify the strong connectivity hypothesis. To this aim, the direction of the edges is eliminated. For the edges that presented both directions, the mean between the two weights is calculated and the result is considered as the new weight for the freshly created undirected arc. The last remark concerns the presence of parallel edges. For those arcs, the classical approach consisting of adding up all the corresponding weights of parallel edges is used [177].

Therefore, two principal manipulations have to be made to the Reconstructed Transition Multigraph $G^M_{1/w_C}(\mathcal{T})$: the substitution of the parallel edges $a^k_{i,j} \in E^M(\mathcal{T})$ with an arc $a_{i,j} \in \hat{E}^M(\mathcal{T})$ presenting as weight the sum of the values of parallel arcs, and the creation of symmetry in the network. Mathematically, these operations leads to the creation of a new set of edges $\hat{E}^M(\mathcal{T})$ and a new weight matrix $\hat{W}^M_{1/w_C}(\mathcal{T})$:

$$\hat{E}^M(\mathcal{T}) = \{a_{i,j} = (s_i, s_j) \in V^M(\mathcal{T}) \times V^M(\mathcal{T}) | \exists a^k_{i,j} \in E^M(\mathcal{T}) \vee \exists a^k_{j,i} \in E^M(\mathcal{T})\},$$

$$(5.56)$$

$$\left(\hat{W}^{M}_{1/w_C}(\mathcal{T})\right)_{i,j} =$$

$$\begin{cases} \frac{\sum_{k=1}^{|\mathbb{A}|}\left(W^{M}_{1/w_C}(\mathcal{T})\right)^{k}_{i,j}+\left(W^{M}_{1/w_C}(\mathcal{T})\right)^{k}_{j,i}}{2}, & \text{if } \exists a^{k}_{i,j}, a^{k}_{j,i} \in E^{M}(\mathcal{T}) \\ \sum_{k=1}^{|\mathbb{A}|}\left(W^{M}_{1/w_C}(\mathcal{T})\right)^{k}_{i,j}+\left(W^{M}_{1/w_C}(\mathcal{T})\right)^{k}_{j,i}, & \text{if } \exists a^{k}_{i,j} \in E^{M}(\mathcal{T}) \oplus \exists a^{k}_{j,i} \in E^{M}(\mathcal{T}) \\ 0, & \text{otherwise} \end{cases}$$

$$(5.57)$$

where the weights are differentiated in three cases: when there are edges pointing in both the directions, then the resulting weight is the average between the sum of both ways weights, when there is only one direction between two nodes (the symbol $\oplus$ indicates the xor operator), then the computed weight is the sum of the parallel edges, and in all the other cases, the considered weight is equal to 0. Therefore, the tensor $W^{M}_{1/w_C}(\mathcal{T})$ is transformed in a matrix $\hat{W}^{M}_{1/w_C}(\mathcal{T}) \in \mathbb{R}^{|V^{M}(\mathcal{T})|\times|V^{M}(\mathcal{T})|}$, which represents the weighted adjacency matrix of the modified Reconstructed Transition Multigraph.

In this way, the modified graph $\hat{G}^{M}_{1/w_C}(\mathcal{T}) = (V^{M}(\mathcal{T}), \hat{E}^{M}(\mathcal{T}), \hat{W}^{M}_{1/w_C}(\mathcal{T}))$ can be obtained. Then, for each connected component of the latter graph, it is possible to compute the eigenvector centrality.

**Theorem 5.7** *Given the modified Reconstructed Transition Multigraph $\hat{G}^{M}_{1/w_C}(\mathcal{T})$, the eigenvector centrality exists and is unique for each of the connected components of $\hat{G}^{M}_{1/w_C}(\mathcal{T})$.*

**Proof.** The proof of this theorem relies principally on the Perron-Frobenius Theorem (Theorem 5.1) and Proposition 5.1.

For each connected component of $\hat{G}^{M}_{1/w_C}(\mathcal{T})$, the condition of the non-negativity of the weights is satisfied, and the strong connectivity is obtained by the restriction of the considered graph and the modifications made in the Reconstructed Transition Multigraph. In this way, the hypotheses of the Perron-Frobenius Theorem (Theorem 5.1) are satisfied. Hence, the result is proved.                               $\square$

With this last theorem, the existence conditions for all the metrics analyzed in the multigraph context are ensured. In the next section, the multilayer version of the Reconstructed Transition Multigraph is defined and the relative centrality measures are adapted for the specific structure introduced.

### 5.4.2   The Reconstructed Transition Multiplex and its Centrality Measures

In this section, a different representation is presented for the Reconstructed Transition Graphs, considering the application of *multilayer networks*. Through this structure, each state can be deeply decomposed by projecting it into specific layers. In this way, the importance of the resulting components can be analyzed independently. Furthermore, a different set of metrics that can be defined based on the classical centrality measures.

The central idea of this approach consists of constructing the reconstructed transition graph as a particular kind of multilayer graph, i.e., the multiplex, to produce the *Reconstructed Transition Multiplex*. Thereby, the only connections between diverse layers link the nodes indicating the same information at different levels. Each of the aforementioned layers is determined by a specific action that can be selected by the agent. Therefore, the total number of layers in the Reconstructed Transition Multiplex is equal to $|\mathbb{A}|$, the cardinality of the action set. Maintaining the usual definition that associates the states of the MDP to the nodes in the graph, it is clear that in the multiplex formulation, the vertices determine a couple of states and actions. In fact, the selected node represents an instantiation of the state variables and its layer indicates the action attached to it. Thereby, while studying this type of graph, the vertices assume the connotation of state-action pairs.

Additionally, in a deterministic environment, there is at most only one outgoing edge each state, since the arc is strictly connected to the action layer where it is lying. Indeed, during the construction process of the Reconstructed Transition Multiplex, every time a new state is discovered, all the layers are updated by adding the states decomposed representations in each layer, while, if an action is been performed for the first time in this freshly encountered state, then an edge, connecting the initial state and the subsequent one, is generated only on the layer corresponding to the chosen action. A complete overview of the creation of the Reconstructed Transition Multiplex is later discussed in detail.

After outlining the general concept utilized, it is now necessary to formally define the Reconstructed Transition Multiplex in order to avoid any possible misunderstandings.

**Definition 5.3** *Given a set of trajectories $\mathcal{T} = \{\tau_p\}_{p=1,\dots,E}$ performed in a MDP $(\mathbb{S}, \mathbb{A}, r, \mathbb{T}, \gamma)$, the Reconstructed Transition Multiplex $G_w^L(\mathcal{T})$ with inter-layer weights $\omega_I$ is defined as follows:*

$$G_w^L = \{G_w^l(\mathcal{T}) = (V^l(\mathcal{T}), E^l(\mathcal{T}), W_w^l(\mathcal{T}))\}_{l=1,\ldots,|\mathbb{A}|} \tag{5.58}$$

*with*

$$V^l(\mathcal{T}) = \bigcup_{p=1}^{E} \mathbb{S}(\tau_p), \qquad\qquad \forall l = 1, \ldots, |\mathbb{A}|, \tag{5.59}$$

$$E^l(\mathcal{T}) = \{a_{i,j}^l = (s_i, a_l, s_j) \in \bigcup_{p=1}^{E} \mathbb{SAS}(\tau_p)\}, \quad \forall l = 1, \ldots, |\mathbb{A}|, \tag{5.60}$$

$$W_w^l(\mathcal{T}) = \left( (W_w^l(\mathcal{T}))_{i,j} \right)_{i,j=1,\ldots,|V^l(\mathcal{T})|} \qquad \forall l = 1, \ldots, |\mathbb{A}|, \tag{5.61}$$

$$(W_w^l(\mathcal{T}))_{i,j} = \begin{cases} w(r(a_{i,j}^l) & \text{if } a_{i,j}^l \in E^l(\mathcal{T}), \\ 0 & \text{otherwise}, \end{cases} \qquad \forall l = 1, \ldots, |\mathbb{A}|, \tag{5.62}$$

*where, $V^l(\mathcal{T})$ denotes the set of the decomposed visited state, $E^l(\mathcal{T})$ represents the set of the performed actions (intra-layer arcs), and $W_w^l(\mathcal{T}) \in \mathbb{R}^{|V^l(\mathcal{T})| \times |V^l(\mathcal{T})|}$ is the matrix of the reward dependent weights defined by the function $w : \mathbb{R} \to \mathbb{R}$, for each layer $l = 1, \ldots, |\mathbb{A}|$.*

In this case, the notation adopted is similar to the one considered for the multigraph approach, with the difference that each state $s_i$ is decomposed for each layer $k$ and it is denoted as $s_i^k \in V^k(\mathcal{T})$. For the edges and weights, instead, the usual notation is applied, i.e., $a_{i,j}^k$ denotes the edge connecting the nodes $s_i^k$ to $s_j^k$ in layer $k$, and $w(r_{i,j}^k)$ indicates the weight associated to the previously identified arc.

With the clarification of the notation employed in the multiplex scenario, it is now possible to discuss how the Reconstructed Transition Multiplex is created. As for their multigraphs versions, also for this task, multiple equivalent procedures can be described: these networks can be generated while the RL agent is learning to solve the MDP by applying Algorithm 5.3 or after having saved the interesting trajectories performed in the specific environment. Considering the latter situation, an example is proposed to elucidate the process of creation of the Reconstructed Transition Multiplex.

---

**Algorithm 5.3** Creation of Reconstructed Transition Multiplex during RL acting [160]

---

**Require:** Number of Episodes $E \in \mathbb{N}$, Policy $\pi$.

1: Initialize empty graphs

$$G_w^l = (V^l, E^l, W_w^l) = (\emptyset, \emptyset, \mathbf{0}), \forall l = 1, \ldots, |\mathbb{A}|$$

2: Initialize empty set of states visited $\mathcal{V} = \emptyset$
3: **for** $episode = 1, \ldots, E$ **do**
4:     Initialize state $s_i$
5:     **if** $s_i \notin \mathcal{V}$ **then**
6:         $\mathcal{V} \leftarrow \mathcal{V} \cup \{s_i\}$                                    ▷ Add vertex $s_i$
7:     **end if**
8:     **for** $t = 0, 1, \ldots, T$ **do**
9:         Take action $a_k = \pi(s_i)$ and observe the next state $s_j$ and reward $r_{i,j}^k$
10:         **if** $s_j \notin \mathcal{V}$ **then**
11:             $\mathcal{V} \leftarrow \mathcal{V} \cup \{s_j\}$                                ▷ Add vertex $s_j$
12:         **end if**
13:         **if** $a_{i,j}^k \notin E^k$ **then**
14:             $E^k \leftarrow E^k \cup \{a_{i,j}^k\}$                      ▷ Add edge from $s_i$ to $s_j$ in layer $k$
15:             Set the associated weight $w(r_{i,j}^k) = w(r(a_{i,j}^k))$
16:             $\left(W_w^k\right)_{i,j} \leftarrow w(r_{i,j}^k)$                        ▷ Update weight $w(r_{i,j}^k)$
17:         **end if**
18:         Update state $s_i \leftarrow s_j$
19:     **end for**
20: **end for**
21: $V^l \leftarrow \mathcal{V}, \quad \forall l = 1, \ldots, |\mathbb{A}|$                      ▷ Update vertex sets
22: **return** $\{G_w^l\}_{l=1,\ldots,|\mathbb{A}|}$.

---

Consider again the example already discussed for the multigraph case, where the trajectories are the following:

$$\mathcal{T}_e = \{\tau_1 = (s_1, a_1, r_{1,2}^1, s_2, a_2, r_{2,3}^2, s_3, a_1, r_{3,4}^1, s_4, a_2, r_{4,3}^2, s_3),$$
$$\tau_2 = (s_1, a_3, r_{1,2}^3, s_2, a_3, r_{2,2}^3, s_2, a_1, r_{2,4}^1, s_4, a_2, r_{4,3}^2, s_3)\}. \tag{5.63}$$

It is possible to find the state set in the same manner as for the Reconstructed Transition Multigraph. So, first, the set of states of each trajectory are computed, $\mathbb{S}(\tau_1) = \{s_1, s_2, s_3, s_4\}$, $\mathbb{S}(\tau_2) = \{s_1, s_2, s_4, s_3\}$. Then, $V^l(\mathcal{T}_e) = \{s_1^l, s_2^l, s_3^l, s_4^l\}$, can be derived $\forall l = 1, 2, 3$. Similarly, the set of edges for each layer can be obtained following the procedure seen in the multigraph example. Specifically,

$$\$A\$ = \$A\$(\tau_1) \cup \$A\$(\tau_1) = \{a_{1,2}^1, a_{2,3}^2, a_{3,4}^1, a_{4,3}^2, a_{1,2}^3, a_{2,2}^3, a_{2,4}^1\} \qquad (5.64)$$

thus, by grouping the edges using the same action, it is possible to find the edge sets for each layer:

$$E^1(\mathcal{T}) = \{a_{1,2}^1, a_{3,4}^1, a_{2,4}^1\}, \qquad (5.65)$$

$$E^2(\mathcal{T}) = \{a_{2,3}^2, a_{4,3}^2\}, \qquad (5.66)$$

$$E^3(\mathcal{T}) = \{a_{1,2}^3, a_{2,2}^3\}, \qquad (5.67)$$

which constitute a partition of the set $\$A\$$. In the same fashion, the set of the associated weights can be calculated using a weight function $f$ and the reward earned in each circumstance:

$$\left(W_w^1(\mathcal{T})\right)_{1,2} = w(r_{1,2}^1), \quad \left(W_w^1(\mathcal{T})\right)_{3,4} = w(r_{3,4}^1), \quad \left(W_w^1(\mathcal{T})\right)_{2,4} = w(r_{2,4}^1)$$
$$(5.68)$$

$$\left(W_w^2(\mathcal{T})\right)_{2,3} = w(r_{2,3}^2), \quad \left(W_w^2(\mathcal{T})\right)_{4,3} = w(r_{4,3}^2), \qquad (5.69)$$

$$\left(W_w^3(\mathcal{T})\right)_{1,2} = w(r_{1,2}^3), \quad \left(W_w^3(\mathcal{T})\right)_{2,2} = w(r_{2,2}^3). \qquad (5.70)$$

and the remaining values of the components of $W_w^1(\mathcal{T})$, $W_w^2(\mathcal{T})$, and $W_w^3(\mathcal{T})$ are set to 0.

With this pipeline, the graphs $G^l(\mathcal{T}_e) = (V^l(\mathcal{T}_e), E^l(\mathcal{T}_e), W_w^l(\mathcal{T}_e))$ are produced and through their composition it is generated the overall Reconstructed Transition Multiplex shown in Fig. 5.11.

The relationship between multigraphs and multiplexes is very strict. In particular, for the context of the Reconstructed Transition Graph, it can be noted that the majority of the information required for building both networks is the same. Additionally, it is possible to straightforwardly construct the multiplex from the multigraph version and vice versa. Indeed, the data stored in the corresponding vertex, edge and weight sets can be translated into this special structure. Hence, the Reconstructed Transition Multiplex is the decomposed representation of the Reconstructed Transition Multigraph: if the graphs at each layer for the former network are unified, then it can be recognised as the multigraph version. However, this flattening is possible only if any information is not lost. A major aspect that has not been addressed so far concerns the inter-layer connections that link the different decomposed states (together). To not have any loss in the data connected to the inter-layer edges, it is necessary to set their weights equal to zero. Thereby, through this classical

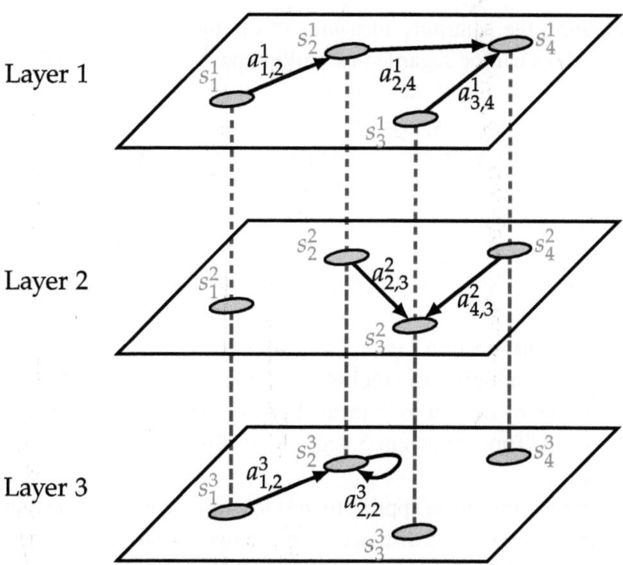


**Fig. 5.11** Example of Reconstructed Transition Multiplex

manipulation, it can be operated equivalently on the collapsed multiplex or the multigraph [27, 49].

Despite the two aforementioned structures modelling the same problem, the adoption of the Reconstructed Transition Multiplex can provide deeper insights into the importance of each state. This facet is conveyed by the association of the state with a specific action, leading in this way to the evaluation of this couple of information. An interesting parallelism can be recognized between the V- and Q-value functions in the MDPs (see Sect. 2.2) and the centrality metrics computed in the Reconstructed Transition Multigraphs and Multiplexes. It can be recalled that while the V-function computes the value of the states, i.e., the maximum of the Q-values, thus, considering the best-case scenario for the specific state, the Q-function analyzes the values linked to the different combinations of state and actions, leading to collecting broader information. In the same manner, the multiplex structure provides a comprehensive knowledge of the vertexes at different levels, whereas the multigraph representation only summarizes the general understanding of the state. Therefore, calculating the centrality measures in the multiplexes will produce a value that can be, theoretically, linked to a decomposition of the same metrics computed in the relative multigraph.

For computing the centrality measures in the multiplexes, the fact that these multilayer versions can be regarded as multigraphs can be exploited. Hence, it is possible to take advantage of the equivalence between the two graphs for the definition of the weights for the specific centrality measures. In this way, the existence conditions found in the previous section can be directly applied in these settings. However, it is fundamental to pay attention to the hypothesis of the theorems. In the following, a detailed analysis of the different requirements needed for each metric is reported.

Starting with the betweenness centrality, it is helpful to recall that the function applied for the calculation of the weights for the multigraphs is $w(r) = -r$. In this situation, this weight function can be used without any other conditions due to the formulation of the multiplex as a multigraph. In fact, the weights of the inter-layer connections are set by default to 0, meaning that the shortest path in both types of graphs is identical. Thus, Theorem 5.4 also holds for the Reconstructed Transition Multiplexes.

A similar procedure can be applied to the closeness centrality. Nevertheless, it is crucial to bear in mind that the distances of the networks have to be strictly positive. So, the inter-layer weights have to be modified to allow the calculation of the length of the shortest path, otherwise, an infinite value is computed. To this end, the inter-layer weights have been fixed at the constant value $\epsilon_I > 0$ such that it is sufficiently large to not massively influence the calculation of the overall closeness measure.

From these remarks, it is clear that the betweenness and closeness metrics can be applied straight to each layer graph. In this way, the problem of the manipulation of the inter-layer values is avoided. On the other hand, this analysis of each decomposed graph can only provide limited details on the overall structure since these projected graphs are sparse. As already mentioned at the beginning of the section, in the deterministic MDPs there is at most one outgoing edge at each vertex in the corresponding layer. So, there is a high number of isolated nodes. For this reason, it has been opted to define the inter-layer weights as $\epsilon_I$ to maintain as much data collected in the MDPs as possible.

However, the particularity of the highlighted structure helps ensure the existence of the closeness metrics in the multiplex version of the reconstructed transition graphs. Indeed, from Theorem 5.5 and Theorem 5.6, it can be noted how the most stringent hypothesis concerns the existence (and, for the former one, uniqueness) of edges with an associated positive reward. These specific conditions are trivially satisfied thanks to the sparse quality of the multiplex generated. Thus, the well-definiteness of the closeness centrality is satisfied by calculating the intra-layer weights through the weight function $w_C$.

The situation gets more complicated in the study of the eigenvector centrality measures applied to the multiplexes. As reported in the preliminary section of the chapter, a wide range of methodologies that compute eigenvector measures can be considered. In this work, three main approaches have been analyzed: the uniform eigenvector centrality [226], the versatility (fourth-order) [57, 58], and the $f$−eigenvector centrality (third-order) [241]. As settled for the other centrality metrics, also in this case the usual weight function $\frac{1}{w_C(\cdot)}$ adopted in the multigraphs is applied for the calculation of the intra-layer weights. Then, a thorough examination of the assumptions for each eigenvector measure is discussed below.

The first considered metric is the uniform eigenvector centrality. This particular centrality measure consists of calculating the classical eigenvector metric using as the adjacency matrix the sum of the adjacency matrices of each layer. Hence, the final computation leads to the same values attained by the multigraphs. Nevertheless, it is fundamental to guarantee that the summed-up matrix satisfies the irreducibility and non-negativity of the elements hypothesis. Only under these assumptions, it is possible to apply the Perron-Frobenius Theorem (Theorem 5.1). However, this approach does not give any additional information that was not already known through the study of the multigraph.

A different outcome is instead obtained by the application of the versatility measure. As seen in Sect. 5.1.2, this metric can be calculated by the evaluation of the eigenvector associated with the largest positive eigenvalue of the supra-adjacency matrix of the multiplex. Despite the simplicity of its computation, the critical assumption requested for the existence and uniqueness of the values found is the strong connectivity of the overall graph. Thus, as done in the multigraph scenario, a manipulation of the multiplex structure has to be operated. In detail, the supra-adjacency matrix used for this calculation is as follows:

$$\hat{B}_{1/w_C}(\mathcal{T}) = \begin{bmatrix} \hat{A}^1_{1/w_C}(\mathcal{T}) & \frac{1}{\epsilon_I}I_{|V|} & \cdots & \frac{1}{\epsilon_I}I_{|V|} \\ \frac{1}{\epsilon_I}I_{|V|} & \hat{A}^2_{1/w_C}(\mathcal{T}) & \ddots & \vdots \\ \vdots & & \ddots & \frac{1}{\epsilon_I}I_{|V|} \\ \frac{1}{\epsilon_I}I_{|V|} & \cdots & \frac{1}{\epsilon_I}I_{|V|} & \hat{A}^{|L|}_{1/w_C}(\mathcal{T}) \end{bmatrix}, \tag{5.71}$$

where $\hat{A}^j_{1/w_C}(\mathcal{T}) = \dfrac{W^j_{1/w_C}(\mathcal{T})+(W^j_{1/w_C}(\mathcal{T}))^T}{2}, \forall j = 1,\ldots,|\mathbb{A}|$, is the average between the weighted adjacency matrix $W^j_{1/w_C}(\mathcal{T})$ of the projected graph at layer $j$ and its transpose $(W^j_{1/w_C}(\mathcal{T}))^T$. With an abuse of notation, it is denoted the cardinality of $V^l(\mathcal{T})$ as $|V|$ in order to simplify the notation used in the definition of the modified supra-adjacency matrix $\hat{B}_{1/w_C}(\mathcal{T})$. Finally, the edge sets $E^l(\mathcal{T})$, $\forall l = 1, \ldots, |\mathbb{A}|$ can be reduced into their undirected version $\hat{E}^l(\mathcal{T})$, $\forall l = 1, \ldots, |\mathbb{A}|$ defined as follows:

$$\hat{E}^l(\mathcal{T}) = \{a_{i,j}^l = (s_i, a_l, s_j) \in V(\mathcal{T}) \times \mathbb{A} \times V(\mathcal{T}) | \exists a_{i,j}^l \in E^l(\mathcal{T}) \vee \exists a_{j,i}^l \in E^l(\mathcal{T})\} \tag{5.72}$$

In this way, all the projected graphs are symmetric (and undirected). This aspect, combined with the use of the inter-layer constant weight $\frac{1}{\epsilon_I}$, ensures the strong connectivity of the multiplex. Hence, the modified Reconstructed Transition Multiplex can be defined as follows

$$\hat{G}_{1/w_C}^L(\mathcal{T}) = \{(V(\mathcal{T}), \hat{E}^l(\mathcal{T}), \hat{A}_{1/w_C}^j)\}_{l=1,\ldots,|\mathbb{A}|} \tag{5.73}$$

With the particular formulation utilized, there is no loss of information for the parallel edges, as instead happened for the multigraph evaluation. For the existence conditions, the following result can be stated as consequence of the Perron-Frobenius Theorem (Theorem 5.1):

**Theorem 5.8** *Given a set of trajectories $\mathcal{T} = \{\tau_p\}_{p=1,\ldots,E}$ performed in a MDP $(\mathbb{S}, \mathbb{A}, r, \mathbb{T}, \gamma)$, the versatility centrality for the modified Reconstructed Transition Multiplex $\hat{G}_{1/w_C}^L(\mathcal{T})$ with inter-layer weights $\frac{1}{\epsilon_I}$ exists and is unique.*

**Proof.** For proving the existence and uniqueness of the versatility centrality it is required to satisfy the hypothesis of the Perron-Frobenius Theorem (Theorem 5.1) for the supra-adjacency matrix associated to the modified Reconstructed Transition Multiplex $\hat{G}_{1/w_C}^L(\mathcal{T})$. In detail, due to the changes made in the definition of the weights and in the general structure of the edges, $\hat{G}_{1/w_C}^L(\mathcal{T})$ satisfies the strong connectivity condition and the supra-adjacency matrix $\hat{B}_{1/w_C}(\mathcal{T})$ present only non-negative values. Hence, the conditions for applying the Perron-Frobenius Theorem (Theorem 5.1) are verified. $\qquad\square$

Despite the versatility metric can extend and improve the classical calculation of the eigenvector centrality for the multigraphs, it still requires strong hypotheses. Thereby, the remaining approach deals with the calculation of the same kind of measurements but with the necessity of weaker requirements.

The $f$−eigenvector metric exists, is non-negative and is unique under the condition that the weights are non-negative (Theorem 5.3). Moreover, it is possible to attain the strict positiveness of the solution by ensuring the existence of an edge with a positive weight for each layer and an outgoing arc with a positive weight for each node (in any of its decomposed vertex). While the former condition is trivially satisfied in the Reconstructed Transition Multiplex, the latter one is not achieved for the terminal states. In fact, for these states, there are no outgoing edges in any layer. However, given the limited amount of terminal states, it is possible to introduce a self-loop for these with a positive weight equal to $\frac{1}{\epsilon_I}$. In this way, the influence generated by this modification has the same order as in the versatility scenario.

Mathematically, these modifications lead to the definition of the new edge sets $\bar{E}^l(\mathcal{T})$ and the new weights $\bar{W}^l_{1/w_C}(\mathcal{T})$ as follows:

$$\bar{E}^l(\mathcal{T}) = E^l(\mathcal{T}) \cup \{a^l_{i,i} = (s_i, a_l, s_i) \in V(\mathcal{T}) \times \mathbb{A} \times V(\mathcal{T}) | s_i \text{ is a}$$
$$\text{terminal state}\} \tag{5.74}$$

$$\bar{W}^l_{1/w_C}(\mathcal{T}) = \left( \left( \bar{W}^l_{1/w_C}(\mathcal{T}) \right)_{i,j} \right)_{i,j=1,\dots,|V^l(\mathcal{T})|} \tag{5.75}$$

with

$$\left( \bar{W}^l_{1/w_C}(\mathcal{T}) \right)_{i,j} = \begin{cases} \left( W^l_{1/w_C}(\mathcal{T}) \right)_{i,j} & \text{if } a^l_{i,j} \in E^l(\mathcal{T}) \\ \frac{1}{\epsilon_I} & \text{if } a^l_{i,j} \in \bar{E}^l(\mathcal{T}) \setminus E^l(\mathcal{T}) \end{cases} . \tag{5.76}$$

Hence, the Reconstructed Transition Multiplex addressed for the $f$−eigenvector centrality is defined in the following manner:

$$\bar{G}^L_{1/w_C}(\mathcal{T}) = \{(V(\mathcal{T}), \bar{E}^l(\mathcal{T}), \bar{W}^l_{1/w_C}(\mathcal{T})\}_{l=1,\dots,|\mathbb{A}|}. \tag{5.77}$$

Despite this insignificant change performed in the multiplex, the general hypotheses assumed are weaker than in the previous contexts.

**Theorem 5.9** *Given a set of trajectories $\mathcal{T} = \{\tau_p\}_{p=1,\dots,E}$ performed in a MDP* $(\mathbb{S}, \mathbb{A}, r, \mathbb{T}, \gamma)$, *the $f$−eigenvector centrality for the modified Reconstructed Transition Multiplex $\bar{G}^L_{1/w_C}(\mathcal{T})$ with inter-layer weights $\frac{1}{\epsilon_I}$ exists and is unique.*

**Proof.** The tensor of the weighted adjacency matrices can be constructed as
$$\bar{W}^{L}_{1/w_C}(\mathcal{T}) = \left( \left( \bar{W}^{l}_{1/w_C}(\mathcal{T}) \right)_{i,j} \right)^{l=1,\ldots,|\mathbb{A}|}_{i,j=1,\ldots,|V^{l}(\mathcal{T})|}, \text{ and it can be noted that it satis-}$$
fies the condition for Theorem 5.3. Thus, the $f$−eigenvector centrality is well-posed. $\qquad\square$

This simplification of the necessity conditions leads to a drawback: the $f$−eigenvector centrality that can be computed indicates the importance measure of the overall state, and not of its decomposed vertices. Hence, it can be only compared with the values obtained by the agglomerated nodes.

After a thorough examination of the metrics proposed for the Reconstructed Transition Multigraphs and Multiplexes, the practical results of their application in a custom environment created for this specific occasion are shown.

### 5.4.3   Computational Experiments

For the computational evaluation of the methodologies introduced in the previous section, a custom environment has been selected for the testing phase: the Taxi with traffic (Appendix A.6 in the Electronic Supplementary Material). In detail, this specific environment is designed in order to satisfy the assumptions for the existence of the different centrality measures. Specifically, a non-sparse reward function is utilized, and multiple critical situations where performing a wrong action can lead to a major loss are introduced. For these reasons, the Taxi environment was a perfect match. However, in order to complicate the task, an addition is made: in some fixed locations, some other cars have appeared, simulating traffic congestion. Hence the custom Taxi with traffic environment is considered in the sequel. The rendering of the map with the trafficked positions is shown in Fig. 5.12, where the other cars indicate the congestion.

For the solution of this problem, a Q-learning agent is trained for $20'000$ episodes and the observed trajectories are stored for the creation of the reconstructed transition graphs. The analysis of the obtained results through the application of the centrality metrics adapted in this chapter is structured in the following manner: first, the general graph information and parameters settings are stated for each scenario, then results attained by the well-known centrality measures from the literature are computed in order to set a baseline. Then, the outcomes of the multigraph procedure are described and, subsequently, the focus is moved to the multiplex version and

**Fig. 5.12** Rendering of the custom Taxi environment with traffic

its results. The aforementioned methodologies are applied to the Taxi with traffic problem for a particular instance: when the passenger is in the bottom-left corner and the destination is in the bottom-right part. For the sake of clarity, the study is limited to the situation shown in Fig. 5.12. Due to the symmetry of the MDP, similar results are produced in the other combinations of passenger-destination locations. Thus, it is sufficient to study this single case.

A final remark, before beginning with the discussion of the computational results, concerns the visualization of the measures: in all the cases, the generated heatmaps present the centrality metrics scaled to the interval [0, 1] to facilitate the comparison between different centrality metrics. In this way, the performances of each method can be discriminated straightforwardly.

## Reconstructed Transition Multigraph Results

For the first part of this computational evaluation, three Reconstructed Transition Multigraphs are created for each of the three centrality measures considered. The generated graphs consist of 404 nodes and $2'400$ edges, including the parallel and the self-loops. For the closeness and eigenvector measures, $\epsilon_C$ is set to 0.1. The choice of this constant is based on a simple analysis of the rewards in the environment. Since the closest reward to 0 is $-1$, it has been decided that a positive action should be at least $\frac{1}{10}$ of the maximum negative value. Further investigations are needed in the future to understand the best settings for each MDP.

The description of the findings can begin from the observation of the heatmaps shown in Fig. 5.13, depicting the importance of each position before picking up the client. Specifically, the first row, i.e., Fig. 5.13a–5.13b–5.13c, reports the results obtained by applying the standard unweighted centrality measures to the simple reconstructed transition graph. These procedures help recognise the topological structure of the environment but without providing any understanding of the map: only the connections between the states are evaluated, leading to not differentiating the trafficked spots from the normal ones. This division has been instead achieved by the employment of the weighted centrality measures for the multigraphs, as represented in Fig. 5.13d–5.13e–5.13f.

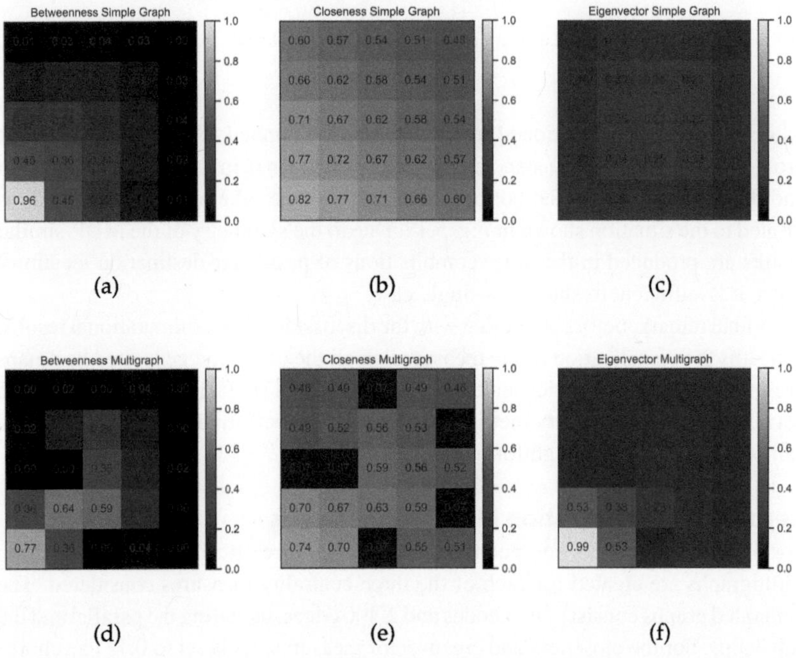

**Fig. 5.13** Heatmaps of the centrality measures in the simple (unweighted) Reconstructed Transition Graphs and Reconstructed Transition Multigraphs for different positions before picking up the passenger

These latter measures show a remarkable ability to identify the passenger position as highly relevant. From the betweenness centrality, the neighboring states around the client are highlighted due to the numerous optimal trajectories passing through them. In particular, a bottleneck caused by the presence of traffic in the nearby sites is identified. Similar results are achieved by the eigenvector centrality, which correctly marks the customer spot and its close states as important. However, the best performances are attained by the closeness measurements. In this case, a complete overview of the taxi map is revealed. In this scenario, not only the traffic zones near the passenger position are correctly spotted, but all the different problematic places are negatively evidenced. Thereby, the performance achieved by the closeness metric outperform for this problem the other methodologies. This result is achieved due to the adoption of the rewards as distances, allowing the closeness measurements to have a clearer characterization of each state's positioning. In fact, in contrast to the betweenness centrality which exclusively considers the number of the shortest paths passing through a node, the closeness measure provides more information thanks to the calculation of the proper distances between each pair of vertices.

The quality of the multigraphs centrality metrics is also confirmed in the second part of the task: after picking up the customer. The heatmaps showing the values computed are portrayed in Fig. 5.14. The same comment can be stated for the simple unweighted measures: since no data regarding the reward function has been taken into account, there is just a representation of how the nodes are connected. Meanwhile, incorporating the rewards in the structure used for the Reconstructed Transition Multigraphs led to the extension of the intuitions to critical states and highly visited nodes. Nevertheless, the outcomes observed in this phase, present some pitfalls. Specifically, the betweenness and closeness measures, represented in Fig. 5.14d and Fig. 5.14e respectively, are associated with the central-map vertices higher importance since they are visited more often by optimal trajectories and are in the centre of the graph, minimizing in this way the distances to multiple nodes. Nonetheless, this facet is backfiring due to the impossibility of recognizing the fundamental states related to the goal: the destination position is not highlighted in either case. On the contrary, the destination is recognized as one of the states having the smallest importance. This particular outcome is generated by the standard definition of these two centrality metrics which award higher values to states that occupy a central position in the trajectories, hence the ones that are often visited, leading to diminish the relevance of the final states. Despite not being able to characterize the goal states, the betweenness and closeness metrics can indicate both the physical bottlenecks determined by the map topology and the risky sites with penalty rewards.

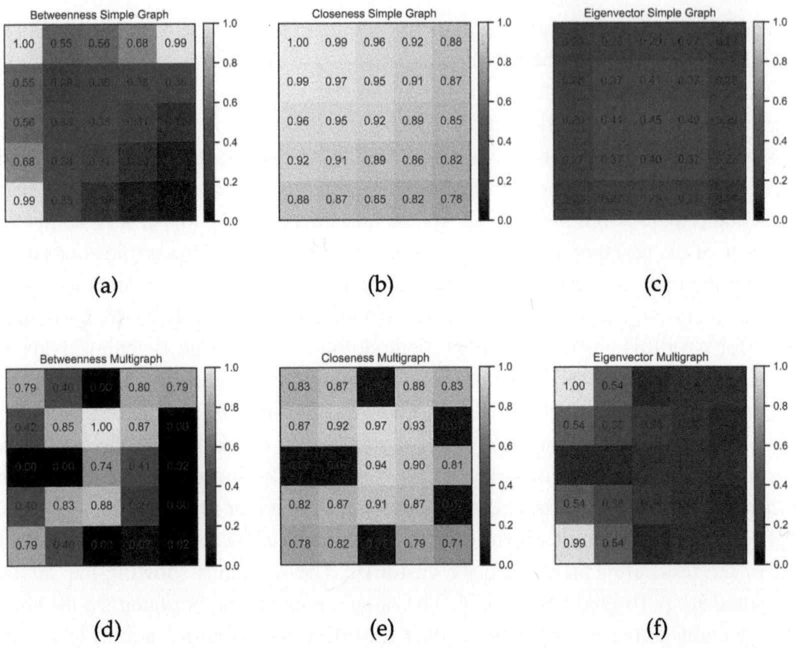

**Fig. 5.14** Heatmaps of the centrality measures in the simple (unweighted) Reconstructed Transition Graphs and Reconstructed Transition Multigraphs for different positions after picking up the passenger

A different result is instead shown in Fig. 5.14f, where the eigenvector centrality is used. In this scenario, the influence of the modifications made to the Reconstructed Transition Multigraph in order to satisfy the existence and uniqueness conditions can be quantified. Indeed, the importance values computed are corrupted by the major changes in both the structure of the graph and the weighting of the edges. Hence, not only the values are distorted but also the relative positions in the map seem to be mixed. This particular aspect of the eigenvector centrality measurements is present in the multiplex calculations as well. However, through the adoption of an approach requiring less stringent assumptions, the computational results can be improved.

## Reconstructed Transition Multiplex Results

Moving now to the discussion about the results of the Reconstructed Transition Multiplex, it is first necessary to explicate the parameter settings. The constant $\epsilon_C$ is set to 0.1 fixed as in the previous case, while the weight for the inter-layer connections $\epsilon_I$ is set to 0.01. In this case, due to the decomposition of the states into state-action couples, the heatmaps are grouped for action layers and the corresponding centrality measures computed are shown in the previously fixed scenario of the environment. Then, the agglomerated measurements are calculated for each complete state: all the values corresponding to each decomposed couple of the particular state are summed up and rescaled accordingly to the overall distribution. Thereby, both the decomposed and full perspective on the obtained importance measurements are provided.

As previously done for the multigraph part, the description of the results is split between before and after picking up the passenger. In detail, the outcomes observed in those two circumstances are shown in Fig. 5.15 and Fig. 5.17, respectively. In the former case, it can be noticed that the focus of the betweenness and closeness maps is in the right corner where the client is waiting. However, the division of the state representation using its projections in the action layers creates an issue: while in the bottom-left corner, it is fundamental to operate the pick up action, it is also fundamental to not perform the drop off. Nevertheless, this latter state-action is identified as equally important as for the pick up situation. This peculiar finding is explained by the presence of inter-layer connections with small weights. In this way, it is possible to connect the two decomposed states through just a link. Thereby, the shortest paths that pass through one node can present slightly equal numbers and distances to the wrong couple.

Another interesting remark concerns the elimination of the insights about the trafficked areas in all the maps. In fact, there is a distinct loss of information due to this decoupling effect. This phenomenon is justified by the splitting of the importance between a higher amount of vertices of the graph. Hence, the generated maps do not show anymore any risk-awareness qualities. The aforementioned condition does not change in the visualizations of the agglomerated values, depicted in Fig. 5.16. In this circumstance, the bottom-left corner is still identified as the most important area.

For the versatility measurements in Fig. 5.15c and Fig. 5.16c, the same issue encountered in the evaluation of the eigenvector centrality for the multigraph version is found. Therefore, it is trivial to understand that modifying the structure of the reconstructed transition graphs, in order to allow these metrics to exist and be unique, does break their efficiency. In this case, the majority of the importance calculated using the versatility approach are small positive values which are widespread in a large range of different states. Thus, it is complex to achieve decent results in the discrimination of the proper fundamental nodes. This aspect is analyzed deeply in the next section.

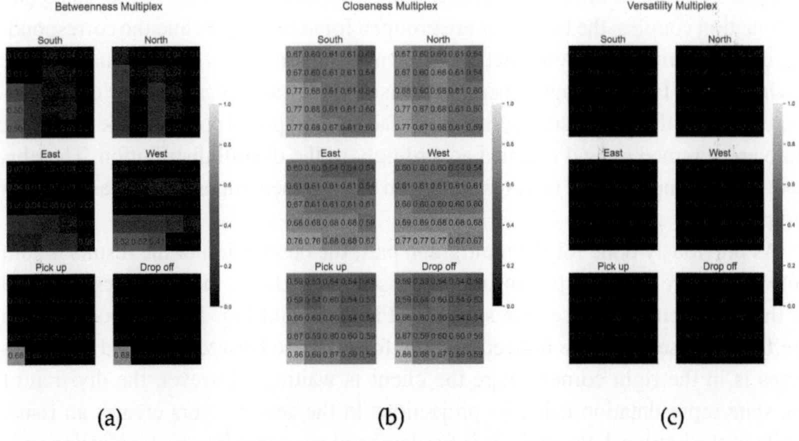

**Fig. 5.15** Heatmaps of the centrality measures in the Reconstructed Transition Multiplex for different positions before picking up the passenger

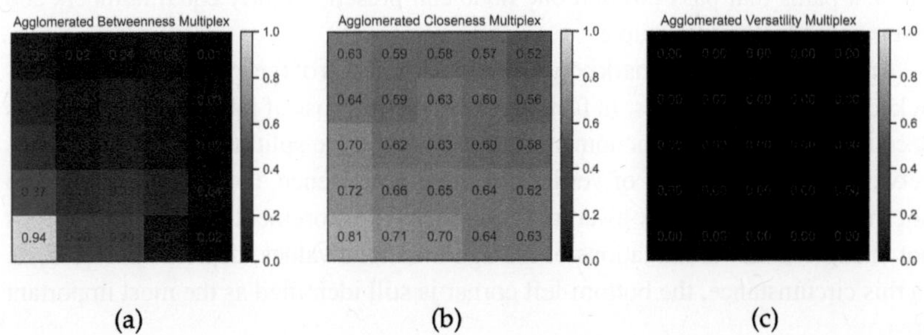

**Fig. 5.16** Heatmaps of the centrality measures in agglomerated states from Fig. 5.15

Similar results are shown in the second phase of the Taxi problem, as represented in the heatmaps of Fig. 5.17 and Fig. 5.18. In these cases, the performances are worse as seen during the analysis of the multigraphs methods. In particular, for the betweenness and closeness centrality, the states that are put in evidence are still the ones in the central part of the trajectories. Moreover, the loss of information about the dangerous zones is perpetuated as in the previous visualizations. Even though the agglomerated values show some insights into the nodes in the centre of the graph, they can not recognize any valid data connected to the goal state. Therefore, based on the results shown the superiority of the multigraphs measurements is clear regarding the identification of the proper important and risky states through the adoption of the betweenness and closeness metrics.

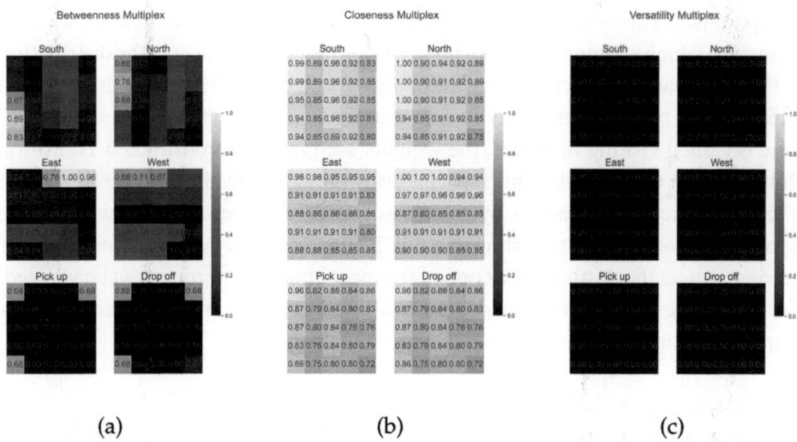

Fig. 5.17 Heatmaps of the centrality measures in the Reconstructed Transition Multiplex for different positions after picking up the passenger

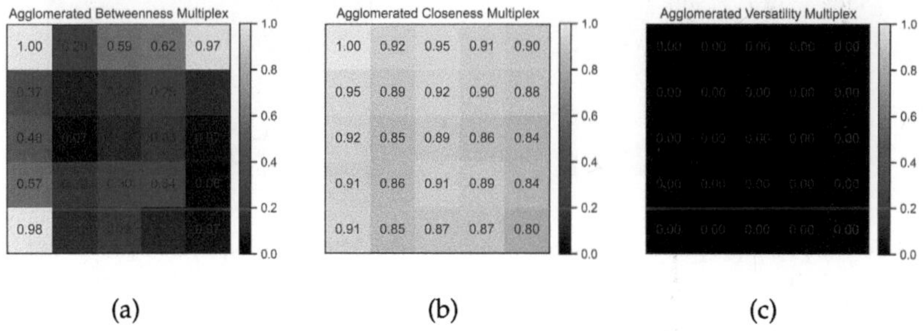

Fig. 5.18 Heatmaps of the centrality measures in agglomerated states from Fig. 5.17

However, this finding can not be claimed for the comparison between the eigenvector measurements. So far neither the classical eigenvector centrality in the multigraphs nor the versatility of the multiplexes has provided satisfying results due to the extensive modification made in the respective graphs. Nevertheless, the outcomes of the last kind of eigenvector centrality have to be discussed.

The results of the computation of the $f$−eigenvector centrality measures using $\alpha = 2.1$, $\beta = 2$, and the convergence threshold at $0.0001$, are shown in Fig. 5.19. In the first part of the problem (Fig. 5.19a), it can be noted that the states that are in the middle of the map achieve larger values thanks to their high connectivity to multiple parts of the graph. Indeed, the corner states are always reaching lower levels of importance. The only exception to this rule is represented by the location with the passenger, which attains a relatively higher value (0.71). This facet in the distribution of the importance is a characteristic of the $f$−eigenvector methodology which combines not only the connectivity of a graph but also its influence in the overall score obtained visiting it. Hence, the weights used in the Reconstructed Transition Multiplex play a major role. This aspect is also highlighted by the correct recognition of the trafficked sites, caused by their associated lower rewards. Looking at the second phase of the Taxi task shown in Fig. 5.19b, the quality of the $f$−eigenvector metric to detect the destination as one of the most important vertices can be verified. The remaining non-trafficked nodes are evaluated as important only based on their connectivity since they are all linked to the same weight.

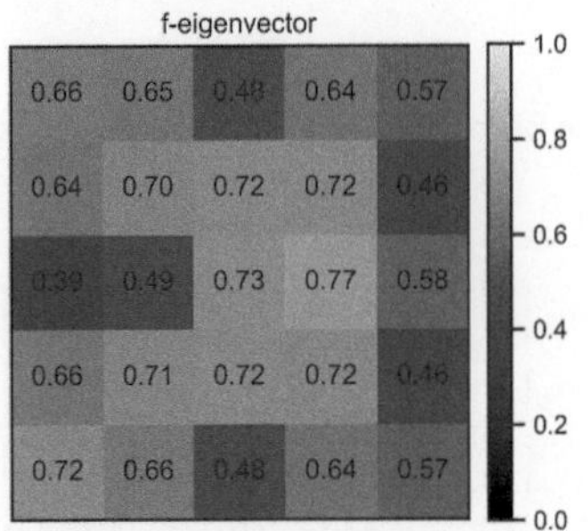

(a) Before picking up the passenger

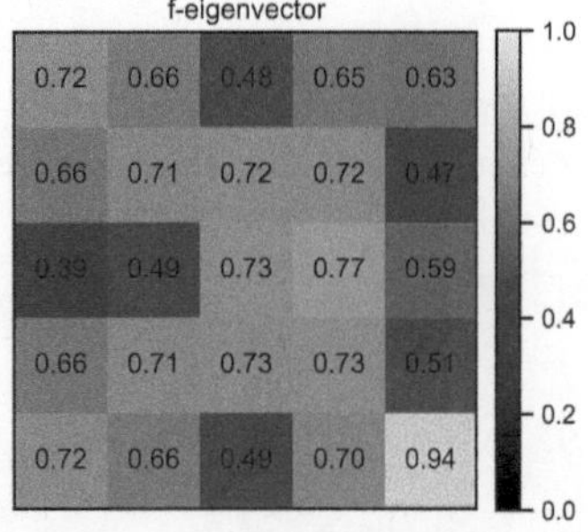

(b) After picking up the passenger

**Fig. 5.19** Heatmaps of the $f$-eigenvector centrality measures in agglomerated states of the Reconstructed Transition Multiplex

The achievements reached by the $f-$eigenvector centrality are by far the most promising ones. Thanks to their limited assumptions, it can give relevant insights into the problem from both the topological and reward point of view. Moreover, it is the only eigenvector metric that is able to create decent results. The explanation of why this methodology is outperforming the one introduced before is addressed in the following section.

### 5.4.4   Discussion

From the results reported in the previous section, it is evident how the multigraph approaches outperform the multiplex versions. The reason principally lies in the high level of decomposition of the states, thus, the discovered importance levels are distributed widely in a large set of nodes. Thereby, there is an increase in the difficulty of recognizing and characterizing the proper important situations. Nevertheless, it is appropriate to investigate the direction of the multiplex centrality metrics due to their ability to analyze the state-action couples as in the Q-function. Hence, specific measures have to be developed in order to highlight only the fundamental decomposed states.

If the vertices with higher betweenness and closeness importance are compared in the two methodological analyses, then it is obvious to claim that the nodes selected in the multigraph case can be interpreted by the definition of the metric; while the ones chosen in the multiplex scenario seem randomly taken. Despite the interpretability capabilities improved through the study of the agglomerated measures, they still can not reach the performances of the multigraph methods. Therefore, it is essential to adapt these metrics to the settings of the Reconstructed Transition Multiplexes to achieve an acceptable level of understandability.

On the other side, the application of the eigenvector centrality metric in the multigraph did not ensure satisfying results in contrast to the clear overview depicted by the $f-$eigenvector centrality applied in the multiplexes. It is useful to compare the distributions of the eigenvector centralities in both experiments in order to have a transparent comprehension of the results shown before. Figure 5.20 illustrates three of these distributions, starting with the eigenvector centrality in multigraphs (Fig. 5.20a), as well as the versatility metrics in multiplexes (Fig. 5.20b), and finally the $f-$eigenvector centrality in multiplexes (Fig. 5.20c). From these plots, it is indeed possible to clarify the reasons why there were a majority of importance values set to 0: the majority of the nodes take values in the neighborhood of 0, and only a few vertices attain higher importance. So, through the linear normalization in the interval $[0, 1]$, just a small set of states maintains a relevant level of centrality.

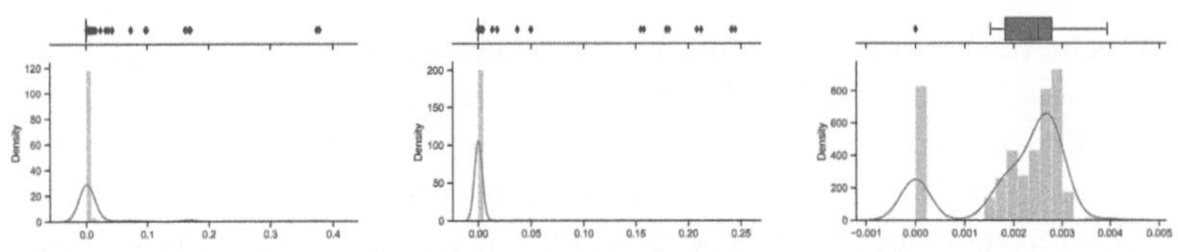

(a) Eigenvector Multigraph   (b) Versatility Multiplex   (c) $f$−eigenvector Multiplex

**Fig. 5.20** Density plot of the eigenvector-related metrics

This situation is only present in the cases when large modifications were made in the Reconstructed Transition Graphs analyzed. In fact, for the $f$−eigenvector centrality, a bi-modal distribution centered in 0 and 0.0025 is discovered. While the first peak corresponds to the states that are not visited because of structural limitations of the environment, e.g., the destination and passenger positions are the same, the second one outlines a valid distribution for the recognition of the important states. By selecting a threshold parameter according to the distribution of the measurements, is indeed possible to discriminate the fundamental states through the adoption of extreme value analysis.

It can be noticed how the statistical analysis can provide helpful insights in the comprehension of the results. Moreover, a critical step in the evaluation of the centrality measures in the network analytics field consists of evaluating the correlation between the different measures utilized. In this way, redundant information obtained by different metrics can be identified. In the next section, the focus is exactly on the comparison between the state-of-the-art methods and the ones proposed in this work.

## 5.5    Comparison of Metrics

The existence of multiple centrality metrics is justified by the fact that there is no unanimity on what the concept of centrality is and there are no agreements in the proper formulation [75]. For this reason, the consequent analysis, that is executed after developing novel methodologies in the network analytics field, consists of evaluating the correlation between the novel approach and the ones in the literature. In this statistical comparison between the multiple importance measures can play a principal role in the discovery of hidden information that was not recognized before [162].

For this reason, the correlation between the measures from both groups (Q-value-based and graph-based) is calculated. In detail, the values used for the correlation study have been computed in the Taxi with traffic environment. A total of 18 metrics were considered in this statistical evaluation, divided into 10 graph-based (betweenness, closeness and eigenvector centrality for the simple unweighted graph, multigraph, and multiplex plus the $f$−eigenvector), and 8 Q-value-based (action gap, advising, mean and difference in both the static and iterated versions).

The choice of mixing metrics from different groups is justified by the possibility of discovering pieces of information that are found in either case. In this way, interchangeable approaches can be identified. Moreover, the Q-value-based methodologies straightforwardly characterize the important states according to a practical definition of the importance. On the other hand, the meaning associated with the important states derived by the graph-based methods is essentially generated by the evaluation of the topological structure of the reconstructed transition graphs. So, they present only a figurative importance that can be interpreted in the MDP context.

Focusing now on the actual results of the correlation analysis, the correlation matrix is reported in Fig. 5.21. The first strong positive correlation that must be noticed is between the betweenness centrality in simple graphs and multiplexes. A similar result is achieved by the closeness centrality. This particular outcome confirms the idea that using the multiplex representation of the reconstructed transition graphs leads to a dispersion of the information. However, this decrease in the data is as extensive as not considering any knowledge of the rewards earned. Hence, this representation heavily limits the viewpoint generated by the multiplex modelling.

Always in the class of the graph-based, it is remarkable how the eigenvector metrics are not correlated to any other approach, the only exception being the $f$−eigenvector centrality, which shows interesting relationships. Specifically, other than the expected high positive correlation with the closeness measures for all kinds of graphs, it presents a medium level ($> 0.50$) of positive correlation with all the static Q-value measurements. Particularly meaningful is the correlation coefficient with the advising and the difference metric, which corroborates the capabilities of the $f$−eigenvector centrality to detect important states in terms of (sub-)goal completion and avoidance of high-risk situations. Therefore, the $f$−eigenvector metric is the only graph-based approach that shows a broad set of information related to the Q-value-based importance values.

In general, a scarcity of correlation between the graph-based and the Q-value-based methods can be observed. This lack of interaction between these two classes is caused by the respective focus points: while the Q-value-based are strictly dependent on the trained agent, the graph-based are partially altered by a different training seed

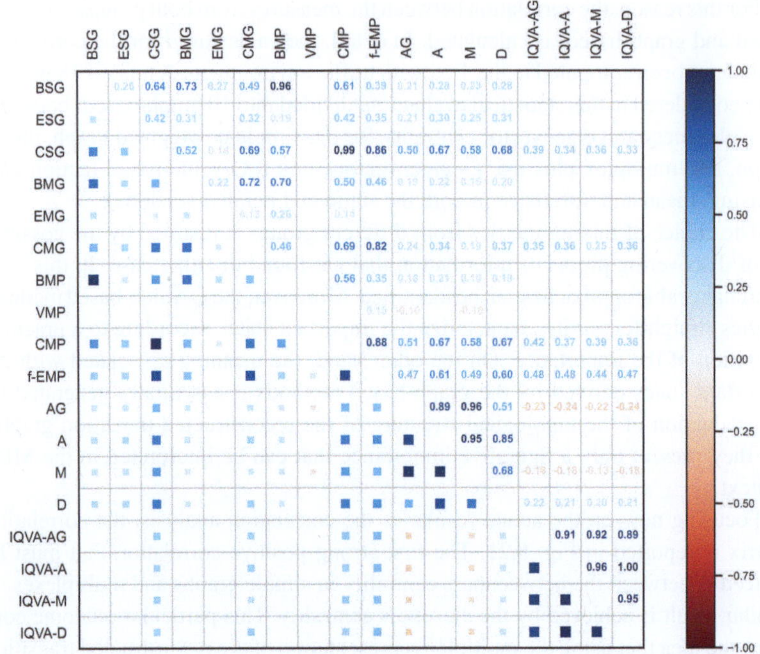

**Fig. 5.21** Correlation matrix of the following measures: betweenness simple graph (BSG), eigenvector simple graph (ESG), closeness simple graph (CSG), betweenness multigraph (BMG), eigenvector multigraph (EMG), closeness multigraph (CMG), betweenness multiplex (BMP), versatility multiplex (VMP), closeness multiplex (CMP), f-eigenvector multiplex (f-EMP), action gap (AG), advising (A), mean (M), difference (D), iterated action gap (IQVA-AG), iterated advising (IQVA-A), iterated mean (IQVA-M), iterated difference (IQVA-D)

or choice in the algorithm used for the solution. In fact, the fundamental data for the methodologies from the latter group consists of the level of exploration in the MDP.

These considerations confirmed how necessary it is to create a specific metric that couples together the information relative to the topological structure of the MDP and the insights gained by a trained agent. To this end, it can be beneficial to adopt the Q-values as weights of the arcs in the reconstructed transition graphs. With this gimmick, it is possible to combine the perks of both the group of methodologies to overcome the previously defined limitations. Furthermore, this statistical analysis has to be extended to a larger set of environments in order to refine the discoveries stated here. Only by studying the outcomes of the approaches from the two distinct classes in a multitude of MDPs, can be discriminated the advantages or the pitfalls of each metric.

## 5.6   Answering Research Question 2

In this chapter, attention has been brought to the metrics helpful in detecting the important states in MDPs. However, the approaches from the literature presented limitations that affect their capabilities in the determination of the distal states. Specifically, the characteristics that are not recognized by the state-of-the-art methodologies involve the combination of the bottleneck properties of the states and their relevance in attaining the maximum return at the end of the episode. Hence, it is fundamental to mix the qualities regarding the structure of the environment and the insights on the reward function to detect the distal states. The methodologies that have been developed in this chapter satisfy these conditions, and especially are able to answer the RQ2:

> **RQ2**: How can the distal state be automatically detected through the adoption of the importance metrics?

An extension to both the Q-value-based and graph-based metrics has been proposed leading to the creation of multiple novel measures able to directly characterize the distal states. Specifically for the Q-value-based approaches, the classical measures, computing the importance measurements as linear combinations of the outputs of the state-action value function, are formulated in an iterated fashion to include also information regarding the state visitation. In this way, the resulting calculations can be associated not only with the final return achieved but also with the frequency of inspection of that situation.

On the other hand, for the graph-based procedures, new theoretical frameworks, extending the usual Reconstructed Transition Graphs, have been formulated. In particular, a multigraph and multiplex version of the aforementioned graphs have been introduced, together with the assumptions required for ensuring the existence and uniqueness of the connected centrality metrics. With these equivalent formulations, the rewards gained along the training phase can be efficiently used in obtaining helpful perceptions on the correct trajectories to follow in the graphs. Thus, also in this case, both the attributes linked to the distal states are discriminated with the analysis of these updated networks.

Nevertheless, improvements in both groups are still required. Indeed, a major issue that has been discussed in the correlation analysis is the lack of a measure that is able to couple together the information regarding the topology of the MDP and the intuitions of the trained agent. The future direction hereby has to address this problem and maybe try to mix the methodologies coming from the two groups of metrics in order to broaden the overview generated by these measurements.

A different question, which will be examined in the successive methodological chapter, concerns the composition of a completely automatic XRL approach merging the importance metrics presented here and the causal explanation generator (BENEDICT) described in Chap. 4. This specific challenge is central to answer RQ3, the last research question.

# Auto-BENEDICT: Combining Importance Metrics and BENEDICT for an Automatic Explanation Generator

**6**

An open question regarding the possibility of completely automating the approach presented in Chap. 4 is addressed in this last methodological chapter. It has already been discussed in the previous chapters how the proposed BENEDICT approach still lacks an autonomous process for the generation of distal information in order to be completely automated. Indeed, the procedure for creating the causal explanations does not require any other updates given the proposed methods applied to the RL scenarios. On the other hand, the component relative to the distal information forecasting necessitates a ground-truth dataset to train the actual and counterfactual RNNs. To this end, the novel metrics introduced in Chap. 5 can be critical in resolving this issue. In particular, they are able to recognize the states that share the main attributes and qualities associated with the distal states. Thereby, after detecting the distal states, the linked actions can be easily identified by looking at the RL policy in those situations. However, it is not trivial to define the pipeline of production for the distal dataset. Moreover, while the methodologies included in Chap. 4 are predominantly practical, the extended metrics have a more sophisticated theoretical background. Hence, the combination of these methodologies can generate complications due to the distance between the two distinct fields.

The proposed approach, already published in Milani [157], is based on the evaluation of the importance metrics for the state trajectories and the subsequent detection of the peak in the obtained time series. Through this idea, the *relevant states* can be recognized. Despite the simplicity of this process, it is first necessary to clearly define how to discriminate the *relevant* information extracted from the stored trajectories through the novel metrics considered. After this preliminary step, the new pipeline combining the previously described BENEDICT method with a metrics-based component can be outlined. In this way, the completely automatic procedure denoted as *Auto-BENEDICT* can be presented and the results associated with its applications



R. Milani, *Advanced Automation for Comprehensible Causal Explanations of Reinforcement Learning Agents*, https://doi.org/10.1007/978-3-658-50495-3_6

are illustrated. Finally, the third research question is answered by summarizing the overall process.

## 6.1    Coupling Importance Metrics and Explanation Generator: the Auto-BENEDICT Approach

In the following, the application of importance metrics is implemented in the context of the analysis of the state trajectories. Here, the main idea consists of evaluating the associated measures for each of the visited states and, by studying the obtained importance time series, extracting the most important moments. In detail, the critical states can be detected in the sequence of observations by inspecting the peaks of the time series. An outlier value in a neighborhood can be indeed recognized as an essential step for the completion of the task or a high-risk circumstance where a careful decision has to be taken. An example is visualized in Fig. 6.1, where three peaks have been marked.

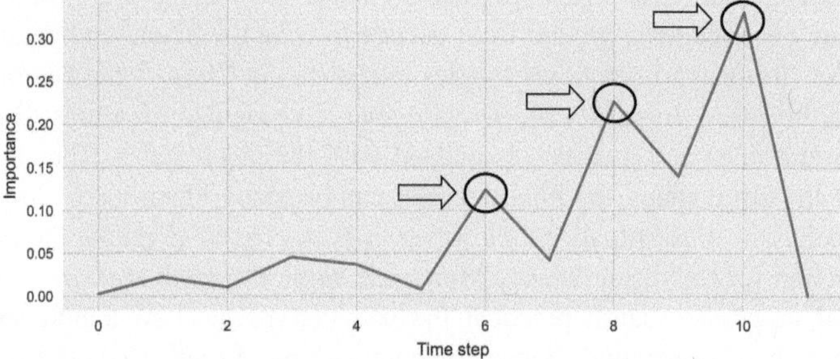

**Fig. 6.1** Importance time series of the visited states with marker peaks associated with the relevant states ©2024 IEEE

However, not all the peaks can be connected to a relevant piece of information. Indeed, in the plot above, two smaller peaks can be noticed before the three identified earlier. These two outliers do not have a significant increment ($> 0.02$) with respect to their neighborhood. The assumption of a confidence interval is crucial for avoiding inaccuracies in the evaluation of non-significant states. For this reason, the difference between the maximum value and the corresponding neighborhood has to be greater than a specific positive threshold. Formally, the *relevant states*, used in the phase of constructing the dataset for the training of the RNN in predicting the distal information, can be defined as follows:

**Definition 6.1** *Given a state trajectory $\tau_s = (s_0, s_1, \ldots, s_{t-1}, s_t, s_{t+1}, \ldots, s_T)$ and an importance metric $I : \mathbb{S} \to \mathbb{R}$, a relevant state $s_t$ in the state trajectory $\tau_s$ is defined such that:*

$$I(s_{t-1}) + \Delta \leq I(s_t), \tag{6.1}$$

$$I(s_{t+1}) + \Delta \leq I(s_t), \tag{6.2}$$

*where $\Delta > 0$ is the threshold for the inclusion of the peak.*

The choice of the threshold parameter $\Delta$ plays a central role in deciding the amount of information considered. When the value is large, a smaller portion of states will be regarded as relevant. Whereas, if it is small it can be generated a large set of states. Hence, it is critical to determine a precise way of characterizing this choice. In detail, it is possible to evaluate always the importance metrics in a normalized manner (with values in $[0, 1]$) and ensure that $\Delta \in [0.01, 0.05]$. This specific choice for $\Delta$ is justified by tracing parallelism between the significance of a statistical test and the significance of the difference in importance between two consecutive states: statistically $0.05$ is considered a reliable threshold for the $p-$value, meaning that the null hypothesis has a 5% chance of being true. Hence, translating this in the context of the importance measures, it is sufficient to have a difference in states that corresponds to a value in-between the 1% and 5% of the maximum measurement. Then, the specific choice of the value inside of the interval can be decided through an accurate analysis of the distributions of the importance values computed for the tackled problem. This latter process is discussed in detail in the computational experiments reported in later parts of this thesis.

Applying the definition above, the target distal information required for the training of the RNNs has been identified. Thus, the process of prediction of the distal state can follow the usual pipeline presented in Chap. 4. The associated distal actions are selected by observing the RL policy: for each relevant state, it is possible to consider the action chosen by the RL agent. Lastly, the final return can be easily computed at the end of the episode by summing up all the rewards obtained at each time step.

Nevertheless, the adoption of the importance metrics can also be considered in terms of evaluating the risk of each instantiation. Indeed, multiple measures are able to discriminate high-risk states, e.g., the difference metric and the closeness centrality. Despite their capabilities, these methodologies have not been considered for the generation of risk-aware statements, so the focus has now shifted in this direction.

Specifically, the computation of the criticality or risk level can be performed by a statistical analysis of the overall risk measurements attained. The risk level is defined by the different quartiles in which the considered importance measure is. Thereby, this data can be included in the explanations generated for both the "Why" and "Why not" questions.

**Definition 6.2**  *Given the BN $B = \{G_B, P_B\}$, a GRU network $f_{GRU}$, a state $s_t$, an actual action $a_t$, the criticality-aware explanation for the "Why action $a_t$?" question is defined as a tuple:*

$$(s_t, \; a_t, \; C(s_t), \; P_{act}, \; d_{act}, \; Pa(D), \; Ch(D), \; R(D)), \tag{6.3}$$

*where $C(s_t)$ is the criticality level of state $s_t$ and the remaining values are the same as in Definition 4.9.*

The textual template used for practically translating the mathematical definition above into a readable statement is an updated version of the already discussed statement in Sect. 4.3.4. The improved template is the following:

"Since the risk is at a *risk level* ($C(s_t)$) and *parent nodes of the actual central variable* ($Pa(D)$) is *instantiation of the parent nodes* ($s_{Pa(D),t}$), it is desirable to do action *actual action* ($a_t$) in order to change *actual central variable* ($D$) from *instantiation of the actual central variable* ($s_{D,t}$) to *prediction of the BN* ($P_{D,act}$) to influence the *children nodes of the actual central variable* ($Ch(D)$), because *rewards nodes* ($R(D)$) are connected to the goal. Following this action, it will be possible to do *distal action* and obtain a final return of *actual prediction of the return* ($d_{act}$))".

In the same way, the formal definition of "Why not" questions can be stated.

**Definition 6.3**  *Given the BN $B = \{G_B, P_B\}$, an actual GRU network $f_{GRU}$, a counterfactual GRU network $\bar{f}_{GRU}$ a state $s_t$, an actual action $a_t$, a counterfactual action $\bar{a}_t$, the criticality-aware explanation for the "Why not action $\bar{a}_t$?" question is defined as a tuple:*

$$(s_t, \; a_t, \; \bar{a}_t, \; C(s_t), \; P_{act}, \; P_{count}, \; d_{act}, \; d_{count}, \; C, \; Pa(D), \; Ch(D), \; R(D)), \tag{6.4}$$

*where $C(s_t)$ is the criticality level of state $s_t$ and the remaining values are the same as in Definition 4.10.*

Similarly as in the answer to "Why" questions, also in this case the risk levels are shown at the beginning of the explanations, leaving unchanged the remaining part of the explanation seen in Sect. 4.3.4:

> "Since the risk is at a *risk level* $(C(s_t))$ and *parent nodes of the actual central variable* $(Pa(D))$ is *instantiation of the parent nodes* $(s_{Pa(D),t})$, it is more desirable to do action *actual action* $(a_t)$ in order to change *actual central variable* $(D)$ from *instantiation of the actual central variable* $(s_{D,t})$ to *actual prediction of the BN* $(P_{D,act})$ instead of changing *counterfactual central variable* $(C)$ from *instantiation of the counterfactual central variable* $(s_{C,t})$ to *counterfactual prediction of the BN* $(P_{C,count})$ doing action *counterfactual action* $(\bar{a}_t)$, to influence *children nodes of the actual central variable* $(Ch(D))$, because *rewards nodes* $(R(D))$ are connected to the goal. Following this action, it will be possible to do *distal actual action* and obtain a final return of *actual prediction of the return* $(d_{act})$, while following the counterfactual action will lead to *distal counterfactual action* and obtain a final return of *counterfactual prediction of the return* $(d_{count})$".

With these data generated by the importance metric component, it is possible to automatically produce causal explanations including the distal information. Therefore, the overall pipeline can be updated by the addition of the relevant state and critical-level evaluation parts. The final procedure, called *Auto-BENEDICT*, is shown in Fig. 6.2. The overall methodology is composed of three main blocks: first, the causal part represented principally by the BN utilized for both the predictions and the creation of the causal explanations, second, a distal section regarding the forecasting of distal actual and counterfactual information through the corresponding RNNs, and, finally, the element introduced in this chapter consisting of the importance metrics evaluation. The latter block has the fundamental role of automating the production of the dataset for the training of the RNNs, leading to the automation of distal information generation. Hence, the Auto-BENEDICT does not necessitate human intervention in any of the three aforementioned blocks.

In the following section, the results achieved in the determination of the importance metrics evaluation are discussed in depth. In particular, the attention is centered on the assessment of the third block validity. In fact, it is the only new method that has been added to the already established BENEDICT pipeline. Hence, the remaining part of the algorithmic procedure will follow the pipeline described in Sect. 4.4.

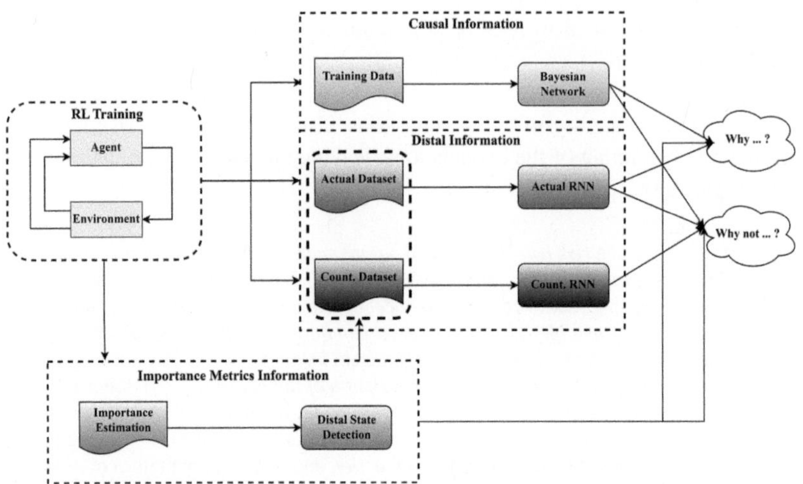


**Fig. 6.2** Pipeline of the *Auto-BENEDICT* approach. ©2024 IEEE

## 6.2 Computational Experiments

For the evaluation of the recognition of the relevant states, and subsequent update of the explanation with the risk levels observed, the classical environment of Taxi is considered. The metrics used in the preliminary phase are twelve (six from the Q-value-based group and six from the graph-based class): Action Gap, Advising, Difference, IQVA-AG, IQVA-A, IQVA-D, and, Betweenness (in Simple Unweighted graphs), Closeness (in Simple Unweighted graphs), Eigenvector (in Simple Unweighted graphs), Betweenness (Multigraph), Closeness (Multigraph), f-eigenvector (Multiplex) centralities. This particular choice is motivated by the requirement of correctly detecting the states that are strictly related to the concept of distal action. Hence, it is fundamental to include in the analysis the proposed measures that should attain this specific task. Moreover, results from state-of-the-art methodologies are presented in order to provide a baseline for the evaluation of the qualities of the newly considered approaches. Specifically, the IQVA measurements were performed after the training of the SARSA agent used for solving the MDP. In the same manner, the static versions of the Q-value-based metrics are computed after the learning phase.

After consistent training of the SARSA method, the Auto-BENEDICT approach can begin with the analysis of the optimal trajectories. To this end, 1000 episodes are completed by the trained RL agent, and the corresponding state instantiations are saved. Thereby, the normalized importance of each state can be calculated by adopting the previously described metrics. The results of the Q-value-based importance time-series analysis are shown in Fig. 6.3, while the achievements of the graph-based metrics are illustrated in Fig. 6.4. In those figures, three subplots are shown for each measurement. In the first column, the mean and standard deviations of the corresponding importance values attained in the stored episodes are reported. In this way, it is possible to grasp the most salient moments in a general trajectory. In the second column, the difference between the actual importance and the previous value is presented. This specific calculation is reported to facilitate the understanding of the selection process for the relevant states. In fact, it must be recalled that the relevant states have the property of presenting a spike in their importance measures. Thus, visualizing where the gap between neighboring (in the time component) states can help the human user to comprehend how the relevant states are spread in the trajectories. Lastly, the frequency of relevant states detected at each time step is represented in the final column histogram. In particular, for the creation of the latter plots, the threshold constant used for recognizing the relevant state has been fixed at $\Delta = 0.02$. This choice is made taking into account the differentiation observed in the second columns of Fig. 6.3–6.4. Indeed, since the mean differences are oscillating around 0 for all the metrics, it has been decided to fix a low value for the determination of the relevant states.

In the following, a more in-depth analysis of the results is discussed. The first observation concerns the comparison between the trends of the importance times series in Fig. 6.3 and Fig. 6.4. In the former case, the trend maintains a simple curvature with a peak in the later stages of the episode. In the latter scenarios, the peak of the mean values is attained in the first part of the trajectories. While this is clear for the betweenness and closeness centralities, a different outcome is delineated for the eigenvector measurements. Furthermore, it can be noticed how the betweenness and closeness centralities in both the simple unweighted graphs and in the multigraphs achieve the same values. The reasons behind this behavior lie in the specific characteristics of the MDP studied: in fact, in the optimal situation, solving the Taxi problem consists of first determining the shortest path connecting the taxi to the passenger, and then, the client to the destination. Therefore, the rewards gained at each time step are equivalent to the distance travelled in the associated graph, meaning that both the formulations are the same. According to this rationale, the discussion of the relative results will be addressed in general as betweenness and closeness measures.

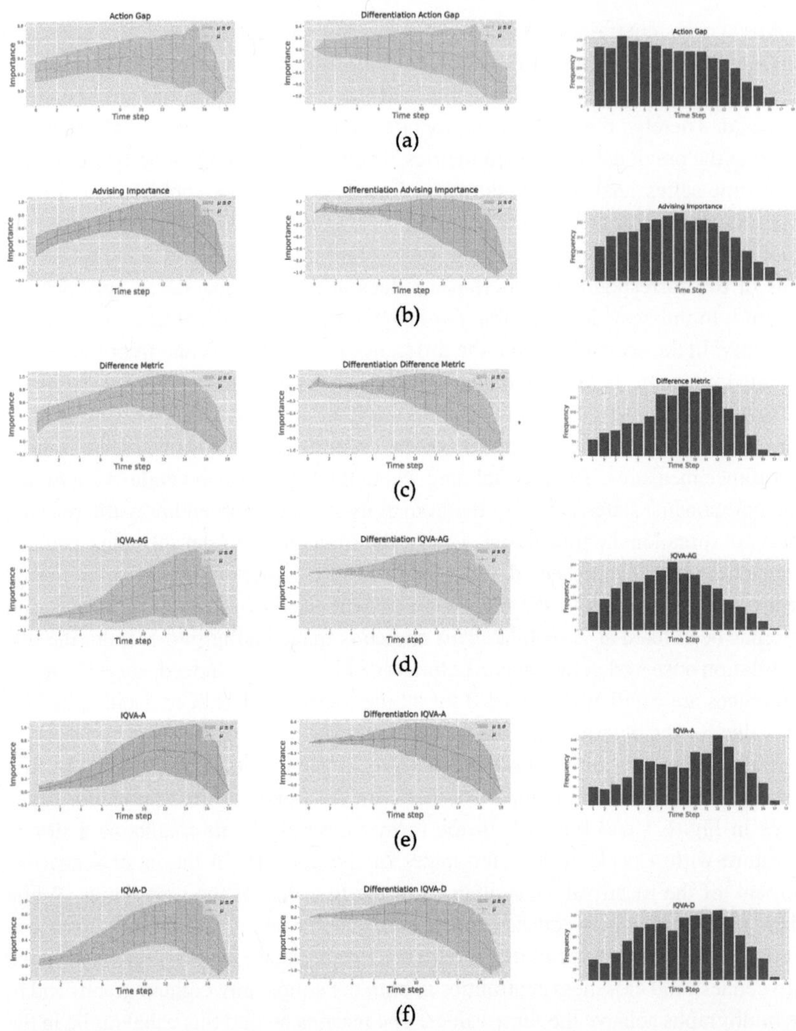

**Fig. 6.3** In the first column, the plots of the average and variance of the importance metrics are reported. In the second column, the plots of the differentiation values with the previous time step can be observed. Lastly, the histograms with the frequencies of the relevant states found using $\Delta = 0.02$ are displayed. The metrics considered are (a) Action Gap, (b) Advising, (c) Difference, (d) IQVA-AG, (e) IQVA-A, (f) IQVA-D. ©2024 IEEE

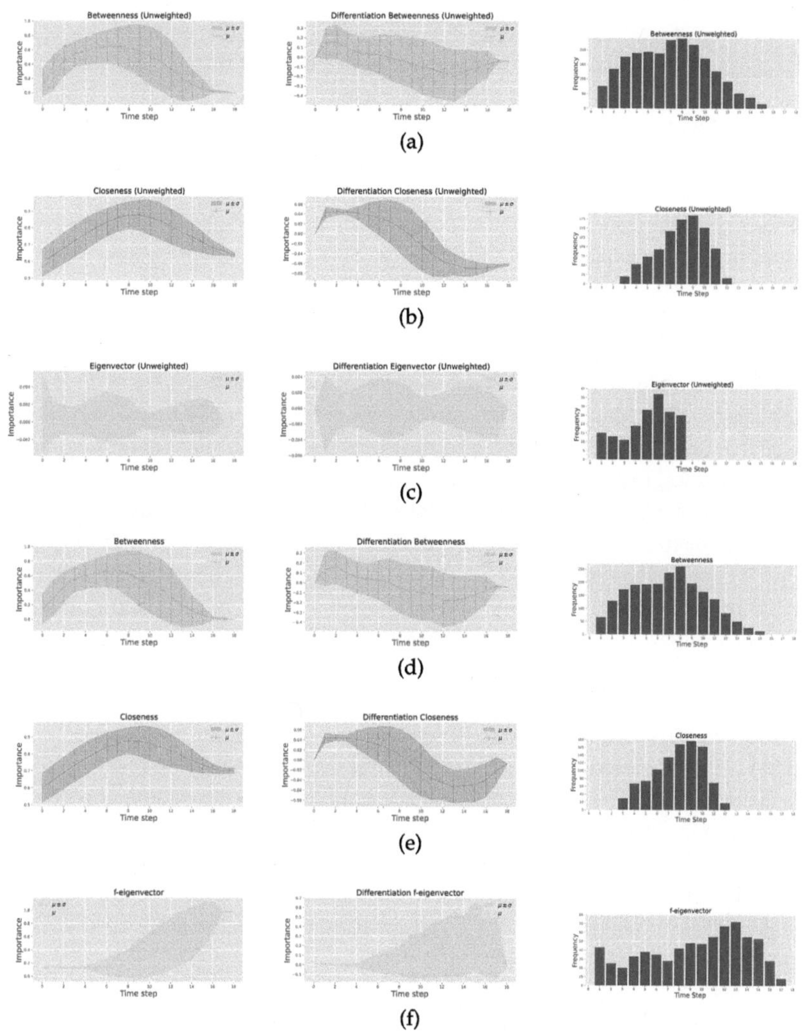

**Fig. 6.4** In the first column, the plots of the average and variance of the importance metrics are reported. In the second column, the plots of the differentiation values with the previous time step can be observed. Lastly, the histograms with the frequencies of the relevant states found using $\Delta = 0.02$ are displayed. The metrics considered are (a) Betweenness (Simple Unweighted Graph), (b) Closeness (Simple Unweighted Graph), (c) Eigenvector (Simple Unweighted Graph), (d) Betweenness (Multigraph), (e) Closeness (Multigraph), (f) f-eigenvector (Multiplex). ©2024 IEEE

An interesting remark can be deduced by looking at the standard deviations of the importance time series. In the static Q-value-based metrics, the variance is maintained constant in the first part of the episode, and then it increases rapidly in the second half. On the other hand, in the iterated measurements, the explosion of the variance happens early in the trajectories, e.g., at $t = 6, 8, 11$ for the IQVA-AG. A similar situation appears in the graph-based measures, where the closeness metric shows a constant standard deviation, whereas the betweenness and $f-$eigenvector values present an expansion of the variance. This phenomenon suggests what the moments are when relevant states are usually visited. Nevertheless, this occasion can also appear later in the episodes, depending on the initialization of the environment, so the variance can be larger in those circumstances.

To corroborate this statement, the distribution of the relevant states found for each metric can be observed in the corresponding histogram. Particularly, it can be noted that the classical Q-value-based measures select more states at the beginning of the episodes, except for the difference metric, which presents an increment of the number of relevant states at the center of the trajectories. On the other hand, the IQVA metrics are generating a bimodal distribution around two central time steps, which are the same where the variance is increasing. This outcome is derived by the property of the IQVA method of assigning higher relevance to states that are visited often. Thereby, the iterated measures outperform the static ones in the discrimination of the correct relevant situations.

Similarly to the latter measurements, the betweenness centrality shows a bimodal distribution in the frequency of the chosen relevant states that can be explained by the same justification used for the IQVA results. Both methodologies take into account the visitation rate of each state in order to determine the corresponding importance. A remarkable outcome is shown in the $f-$eigenvector histogram in Fig. 6.4f. In this case, it is clear how each peak in the number of relevant states corresponds to an increase in the standard deviation at the corresponding moment. Nonetheless, the number of the total relevant states saved is halved if compared to the total amount considered by the IQVA methods.

The number of relevant states and their instantiations are reported in Table 6.1. Here, it is highlighted how the number of relevant states found majorly differs depending on the kind of metric adopted. The methods that are achieving a higher amount of saved data are the static metrics, followed by the IQVA approaches and the graph-based ones. For the eigenvector centralities, the number of states recognized as relevant is significantly lower than in the other scenarios. For the eigenvector centrality in the simple unweighted graph, this circumstance can be justified by the inaccuracies of the evaluations associated with the translation of this metric into the MDP settings. By contrast, the $f-$eigenvector centrality can

**Table 6.1** List of relevant states detected after the analysis of 1000 trajectories considering twelve importance metrics. For each measure, the number of times that the specified state has been individuated as relevant and the particular state instantiations (in order, destination, passenger position, taxi column and taxi row) are reported. ©2024 IEEE

| Top | Action Gap | | Advising | | Difference | | IQVA-AG | | IQVA-A | | IQVA-D | | Betweenness (Unw.) | | Closeness (Unw.) | | Eigenvector (Unw.) | | Betweenness | | Closeness | | $f$−eigenvector | |
|---|---|---|---|---|---|---|---|---|---|---|---|---|---|---|---|---|---|---|---|---|---|---|---|---|
| | # | State | # | State | # | State | # | State | # | State | # | State | # | State | # | State | # | State | # | State | # | State | # | State |
| 1 | 279 | [1, 4, 4, 2] | 270 | [2, 4, 0, 4] | 283 | [1, 4, 4, 1] | 256 | [0, 4, 0, 0] | 273 | [2, 4, 0, 4] | 260 | [1, 4, 4, 2] | 179 | [3, 4, 1, 2] | 164 | [1, 4, 0, 2] | 73 | [1, 4, 3, 4] | 191 | [2, 4, 2, 2] | 179 | [0, 4, 3, 2] | 261 | [3, 4, 3, 4] |
| 2 | 266 | [3, 4, 3, 4] | 264 | [3, 4, 3, 4] | 245 | [2, 4, 0, 3] | 255 | [2, 4, 0, 2] | 254 | [1, 4, 4, 0] | 257 | [0, 4, 0, 0] | 176 | [1, 4, 1, 2] | 157 | [0, 4, 3, 2] | 73 | [1, 4, 4, 3] | 170 | [1, 4, 1, 2] | 174 | [1, 4, 0, 2] | 258 | [2, 4, 0, 4] |
| 3 | 231 | [2, 4, 0, 4] | 234 | [0, 4, 0, 0] | 238 | [3, 4, 3, 3] | 255 | [2, 4, 0, 4] | 254 | [1, 4, 4, 2] | 231 | [3, 4, 3, 3] | 175 | [0, 4, 2, 2] | 104 | [2, 4, 0, 2] | 72 | [2, 4, 3, 4] | 168 | [3, 4, 1, 2] | 99 | [2, 4, 0, 2] | 254 | [1, 4, 4, 0] |
| 4 | 231 | [2, 4, 0, 2] | 232 | [1, 4, 4, 1] | 234 | [0, 4, 0, 1] | 252 | [1, 4, 4, 0] | 241 | [0, 4, 0, 0] | 169 | [2, 4, 2, 2] | 158 | [2, 4, 2, 2] | 91 | [0, 4, 0, 2] | 3 | [1, 3, 4, 4] | 166 | [0, 4, 2, 2] | 87 | [2, 4, 2, 2] | 79 | [3, 0, 0, 1] |
| 5 | 224 | [0, 4, 0, 0] | 186 | [2, 4, 2, 2] | 187 | [1, 4, 3, 2] | 237 | [3, 4, 3, 4] | 232 | [3, 4, 3, 4] | 162 | [3, 4, 1, 2] | 93 | [1, 4, 4, 2] | 90 | [3, 4, 0, 2] | 3 | [3, 1, 4, 4] | 101 | [2, 4, 4, 0] | 87 | [2, 4, 3, 1] | 17 | [2, 0, 1, 2] |
| 6 | 197 | [1, 4, 0, 2] | 173 | [3, 4, 1, 2] | 163 | [2, 4, 1, 2] | 170 | [0, 4, 3, 2] | 188 | [2, 4, 2, 2] | 152 | [1, 4, 0, 2] | 93 | [1, 4, 3, 3] | 87 | [3, 4, 3, 2] | – | – | 101 | [2, 4, 3, 1] | 85 | [3, 4, 3, 2] | 8 | [2, 0, 0, 1] |
| 7 | 197 | [1, 4, 2, 2] | 157 | [1, 4, 2, 2] | 140 | [0, 4, 2, 2] | 170 | [0, 4, 1, 2] | 167 | [1, 4, 0, 2] | 108 | [1, 4, 3, 4] | 89 | [0, 4, 3, 4] | 85 | [2, 4, 3, 2] | – | – | 90 | [2, 4, 3, 4] | 81 | [1, 4, 4, 2] | 8 | [0, 2, 1, 2] |
| 8 | 175 | [3, 4, 1, 2] | 157 | [1, 4, 0, 2] | 96 | [1, 4, 4, 3] | 165 | [1, 4, 2, 2] | 157 | [3, 4, 1, 2] | 89 | [0, 1, 4, 0] | 86 | [0, 4, 4, 0] | 84 | [2, 4, 3, 1] | – | – | 88 | [0, 4, 4, 0] | 81 | [1, 4, 3, 3] | 6 | [0, 1, 4, 2] |
| 9 | 154 | [0, 4, 1, 2] | 153 | [0, 4, 1, 2] | 92 | [1, 4, 0, 3] | 165 | [1, 4, 0, 2] | 151 | [0, 4, 1, 2] | 83 | [3, 0, 0, 0] | 81 | [2, 4, 3, 4] | 84 | [2, 4, 2, 2] | – | – | 86 | [3, 4, 3, 2] | 80 | [3, 4, 1, 2] | 4 | [2, 1, 4, 3] |
| 10 | 103 | [1, 0, 0, 0] | 97 | [2, 4, 3, 4] | 82 | [2, 4, 0, 1] | 91 | [2, 4, 0, 0] | 87 | [1, 4, 3, 4] | 76 | [1, 4, 0, 4] | 80 | [1, 4, 0, 0] | 71 | [1, 4, 3, 3] | – | – | 86 | [3, 4, 4, 0] | 77 | [3, 4, 0, 2] | 3 | [0, 2, 0, 1] |
| 11 | 92 | [1, 4, 0, 4] | 91 | [3, 4, 4, 0] | 77 | [2, 4, 3, 4] | 87 | [3, 4, 0, 0] | 85 | [3, 2, 0, 4] | 75 | [1, 0, 0, 0] | 77 | [0, 4, 0, 2] | 71 | [1, 4, 4, 2] | – | – | 84 | [1, 4, 0, 0] | 70 | [2, 4, 3, 2] | 3 | [1, 0, 3, 0] |
| 12 | 91 | [3, 4, 4, 0] | 88 | [3, 0, 0, 0] | 74 | [0, 4, 3, 1] | 87 | [1, 4, 0, 0] | 81 | [1, 0, 0, 0] | 74 | [2, 0, 0, 0] | 77 | [2, 4, 4, 0] | 67 | [3, 4, 1, 2] | – | – | 78 | [0, 4, 3, 4] | 68 | [0, 4, 0, 2] | 3 | [2, 3, 3, 4] |
| 13 | 86 | [3, 2, 0, 4] | 83 | [0, 4, 3, 1] | 69 | [3, 4, 0, 1] | 86 | [0, 4, 0, 3] | 81 | [2, 0, 0, 0] | 64 | [1, 2, 0, 2] | 77 | [2, 4, 3, 1] | – | – | – | – | 74 | [2, 4, 0, 0] | – | – | 2 | [2, 1, 3, 0] |
| 14 | 84 | [1, 2, 0, 2] | 82 | [1, 4, 0, 4] | 66 | [0, 4, 3, 3] | 84 | [0, 3, 3, 4] | 81 | [1, 4, 0, 4] | 28 | [1, 3, 1, 2] | 73 | [3, 4, 4, 0] | – | – | – | – | 73 | [0, 4, 0, 2] | – | – | – | – |
| 15 | 82 | [1, 4, 3, 4] | 81 | [0, 4, 0, 2] | 65 | [3, 4, 0, 2] | 80 | [0, 4, 3, 0] | 75 | [3, 4, 4, 0] | 5 | [3, 0, 4, 3] | 73 | [3, 4, 3, 2] | – | – | – | – | 72 | [1, 4, 3, 3] | – | – | – | – |

(Continued)

**Table 6.1** (Continued)

| Top | Action Gap | | Advising | | Difference | | IQVA-AG | | IQVA-A | | IQVA-D | | Betweenness (Unw.) | | Closeness (Unw.) | | Eigenvector (Unw.) | | Betweenness | | Closeness | | $f$-eigenvector | |
|---|---|---|---|---|---|---|---|---|---|---|---|---|---|---|---|---|---|---|---|---|---|---|---|---|---|
| # | # | State | # | State | # | State | # | State | # | State | # | State | # | State | # | State | # | State | # | State | # | State | # | State |
| 16 | 82 | [3, 0, 0, 0] | 79 | [0, 1, 4, 0] | 65 | [3, 4, 0, 4] | 80 | [2, 4, 3, 0] | 75 | [0, 1, 4, 0] | 3 | [1, 0, 3, 3] | 72 | [3, 4, 0, 0] | – | – | – | – | 72 | [1, 4, 4, 2] | – | – | – | – |
| 17 | 80 | [2, 4, 2, 2] | 76 | [3, 2, 0, 4] | 59 | [0, 1, 3, 0] | 79 | [3, 2, 0, 4] | 75 | [1, 2, 0, 2] | 2 | [0, 1, 4, 3] | 69 | [2, 4, 0, 0] | – | – | – | – | 66 | [3, 4, 0, 0] | – | – | – | – |
| 18 | 80 | [0, 4, 3, 0] | 75 | [1, 2, 0, 2] | 50 | [1, 3, 2, 2] | 76 | [1, 0, 0, 1] | 63 | [3, 0, 0, 0] | 1 | [3, 1, 1, 0] | 57 | [1, 0, 1, 2] | – | – | – | – | 60 | [0, 1, 3, 1] | – | – | – | – |
| 19 | 80 | [0, 4, 3, 2] | 75 | [1, 4, 3, 4] | 39 | [2, 0, 2, 2] | 74 | [1, 2, 0, 3] | 4 | [1, 0, 3, 3] | – | – | 54 | [3, 0, 1, 2] | – | – | – | – | 55 | [1, 0, 1, 2] | – | – | – | – |
| 20 | 77 | [0, 1, 4, 0] | 71 | [1, 0, 0, 0] | 12 | [1, 3, 3, 1] | 74 | [2, 3, 3, 3] | 1 | [3, 1, 1, 0] | – | – | 51 | [0, 1, 3, 1] | – | – | – | – | 52 | [2, 1, 2, 2] | – | – | – | – |
| Total | 4171 | | 3241 | | 2352 | | 3214 | | 2626 | | 1839 | | 2055 | | 1155 | | 224 | | 2108 | | 1168 | | 906 | |

recognize mainly the destination position as fundamental, and only a few of the other bottleneck positions are also marked as relevant. Due to the limited amount of physical bottlenecks present in the Taxi environment, it is clear that the graph-based measures are not ensuring satisfying performances in the discrimination of the distal states (recognized as the relevant ones).

Turning to the Q-value-based approaches, the one that attains the best performance is the IQVA-AG since it can recognize well-known distal states such as the destination location, the passenger position, and the central-map bottleneck. Moreover, it can also detect the bottlenecks that have to be encountered in order to solve the problem. In fact, on average 3.214 relevant states are identified in an episode, meaning that the two sub-goals relative to the passenger pick up and drop off are always addressed and, at least, an environmental bottleneck is considered. For these reasons, it is selected the IQVA-AG metric for the creation of the actual and counterfactual distal datasets. Specifically, for each of the relevant states, the action, that is followed by the trained RL agent in that situation, is marked as a distal state. Therefore, the distal information that is derived consists of the distal actions that will later be performed by the agent. With this information, the distal part of the explanation can be produced following the pipeline delineated in Sect. 4.3.3.

Satisfying results in terms of correct recognition of the important states (location of the passenger, final destination, and central-map bottleneck) have been obtained by the difference metric and the IQVA-D. However, these measures are principally related to the detection of risky states, so they have been used for the estimation of criticality levels. To this aim, a statistical evaluation of the quartiles for the importance values computed from the stored trajectories is carried out. The box plots shown in Fig. 6.5 depict the different quartiles of each methodology. From this data, it is possible to see the differences between the distributions of the static and iterated measures. Larger values are attained by the classical approach for the first quartile (0.430) and the median (0.613). The gap in the estimations with the IQVA methodology is significant, especially if compared to the first quartile (0.097) and the median (0.205). This aspect can be reasoned by the quality of the IQVA methods of evaluating also the frequency of the states in the overall training. Hence, the states that are inspected by the optimal policy belong to a safer path. For this reason, the second distribution presents lower values which can also be helpful in the identification of more secure criticality bounds. By using the IQVA-D first, second, and third quartile, the number of states indicated as dangerous is higher, leading to a minor probability of misinterpreting an unsafe circumstance.

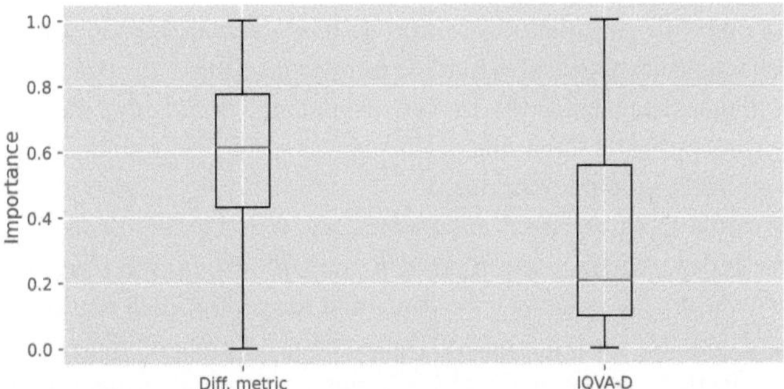

**Fig. 6.5** Box plots of the importance measures associated with risky situations considering the states stored in the analyzed trajectories. ©2024 IEEE

For the creation of the textual answers, the ranges of risk are defined as follows:

- low-risk: when the importance values associated with the state are lower than the median (0.205);
- medium-risk: when the importance values associated with the state are between the median (0.205) and the third quartile (0.556);
- high-risk: when the importance values associated with the state is greater than the third quartile (0.556).

Finally, following the procedure stated in Sect. 4.3.4, the full information can be obtained. Thereby, the explanations provided to the user can be generated by combining the data found from each component. For example, given the instantiation

$$s_t = (\texttt{South-East}, \texttt{South-West}, \texttt{East}, \texttt{North}), \qquad (6.5)$$

and the actual action is $a_t = \texttt{Go West}$, it is required to answer the question "Why action $\texttt{Go West}$?". Using the BN in Fig. 4.9, the next state features are predicted:

$$P_{act} = (\texttt{South-East}, \texttt{South-West}, \texttt{East-Center}, \texttt{North}). \qquad (6.6)$$

Thereby, the central variable $D$ is the taxi column feature. For the distal information, the actual RNN, trained through the previously declared dataset, evaluate the following actual distal information $d_{act} = (\texttt{Go West}, 4)$. Lastly, the risk level has to be estimated by computing the IQVA-D importance measure. So,

$I_{IQVA-D}(s_t) = 0.225$ meaning that the risk level is medium. With these pieces of information, the outcome of the Auto-BENEDICT methodology is the following explanations:

> "Since the risk is at a `medium level` and `Destination` is `South-East`, it is desirable to do action `Go West` in order to change the `Taxi column` from `East` to `East-Center` to influence the `Passenger position`, because `Passenger position` and `Taxi column` are connected to the goal. Following this action, it will be possible to `Go West` and obtain a final return of 4".

In the same manner, the counterfactual questions can be addressed, by including the outputs of their relative components.

## 6.3   Discussion

With the introduction of the importance metric analysis, the initial BENEDICT method has been improved, reaching, theoretically, full independence from human intervention. Nevertheless, this update presents some pitfalls.

The settings of the metrics to adopt and the hyperparameters to set have still to be selected by a human user. In particular, for the former case, it is necessary to have a minimum comprehension of the problem that has to be solved in order to pick the proper importance measure. While, for the latter case, a hyperparameter tuning approach, such as optuna, can be employed, for the evaluation of the importance is fundamental to choose the methodology that is able to highlight the interesting situations. Otherwise, useless or, in the worst-case scenario, counterproductive information will be found.

Additionally, the distal states that have been selected by a specific measure can generate contradictory evaluations between similar trajectories, affecting in this way the learning process and complicating its predictability. Indeed, having a large set of multiple choices will decrease the accuracy of the RNNs in forecasting the appropriate distal information. For this reason, it has been decided to limit the prediction to the distal action in the previous computational experiment.

To simplify this particular forecasting problem, the importance of the nodes of the Reconstructed Transition Multigraph that are in the neighborhood of the actual state can be evaluated. Then, the one achieving the highest value can be related to the next fundamental state. Otherwise, an ad-hoc measure that can be updated according to the actual state visited can be generated. In particular, a metric taking into account the correlation between the present state and a neighboring one can lead to the correct identification of distal states. Moreover, with such an approach, all the pitfalls related to the training procedure of complex RNNs are avoided.

Another aspect that has to be considered is relative to the end goal of the explanations: with the inclusion of the risk level, the users' awareness in deciding on a specific circumstance can be raised. On the other hand, the explanation produced does not provide fundamental information on the action-selection process. So, it can be more beneficial to include data regarding the confidence of the predictions reported. By doing that, the decision-maker can discard unreliable explanations, and personally evaluate the action selected.

Finally, the resulting explanation obtained from the Auto-BENEDICT method has to be validated through a human evaluation study, as done for the BENEDICT case. Only with this test, the improvements that can be achieved with an automatic approach including risk information can be estimated.

## 6.4    Answering Research Question 3

In this chapter, the focus has been turned to the combinations of the two previously defined methodologies. In particular, this coupling has been used to attain autonomy from human intervention. In fact, the central task of this chapter is to find an answer to the remaining RQ3:

> **RQ3**: How can human-validated causal models and the importance metrics be coupled together to improve the automation of the complete explanation-generator pipeline?

In this chapter, it has been remarked that both the BENEDICT methodology and the newly introduced importance metrics can be straightforwardly coupled together. The novel measures are utilized for the discrimination of the distal states, leading to the generation of the actual and counterfactual datasets for the RNNs training. The procedure can be briefly summarized in two steps. First, the state trajectories, stored in the evaluation phase, are transformed into importance time series through the application of the aforementioned measures. Then, the relevant states are identified as the situations associated with a peak in the importance values.

Nevertheless, for the computational experiment, only distal information concerning the distal action has been used. To compute these data, the RL policy is evaluated in the distal states, producing in this way the distal actions. Therefore, the dataset for the training of the RNNs is composed of the distal actions and not of the distal states. This peculiar choice is justified by the difficulty in efficiently predicting this information. Hence, this research direction has to be explored further in future works.

Nonetheless. the importance metrics developed can also be applied to extend the elements provided to the users in the explanation. Specifically, the difference metrics, in both the static and iterated versions, can be used for the evaluation of the risk level of the actual circumstance. Explicitly, through a statistical analysis of the distributions of these importance values in visited states, boundaries for each corresponding criticality level can be determined through the computation of the quartiles. By doing so, the calculated risk can be presented to the human decision-maker, who will take it into account in order to make the best decision given the available data.

The overall pipeline that is created has been defined as Auto-BENEDICT, highlighting its quality of being automatic. With this last methodological improvement, the methodological part of this thesis is concluded. In the next chapter, an application of the proposed approaches to Digital Twins is reported.

# Explainability for Control of Digital Twins: An Application of Auto-BENEDICT

**7**

The concept of Digital Twins has revolutionized the way physical systems are monitored, controlled, and optimized. This technology has seen rapid adoption across industries such as manufacturing [127], healthcare [2], and aerospace [191], where it enables better decision-making, predictive maintenance, and overall system performance improvements. Additionally, these digital replicas can be the perfect environments for training RL agents due to the possibility of simulating different policies without risk. For instance, if the training of the RL agents is performed within the physical systems, then breakdowns caused by wrong choices could occur. Hence, the complications of the training process can be avoided if the RL agents learn the optimal policies in a virtual twin. Moreover, since Digital Twins are applied in particular practical cases, the usage of comprehensible RL agents able to solve specific tasks and produce understandable explanations for the choices selected becomes essential.

In this chapter, the Auto-BENEDICT methodology is applied to a Digital Twin replicating a Robotic Arm. By leveraging the capabilities of the Auto-BENEDICT approach, which combines the automated recognition of relevant states with causal explanations, it is showcased how RL agents can be enhanced for decision-making within complex Digital Twin systems. This integration improves both performance of the RL agent and comprehension of its actions by human users, ensuring trust in the automation process.



R. Milani, *Advanced Automation for Comprehensible Causal Explanations of Reinforcement Learning Agents*,
https://doi.org/10.1007/978-3-658-50495-3_7

## 7.1 Digital Twins

Over the past decades, terms such as cloud computing, platforms, big data, smart cities, ML, and AI have emerged to describe trends in computational and communicative technologies, driving towards automation of specific processes in society [20]. The most recent addition to this list is the concept of "Digital Twin". This idea was firstly considered nearly 60 years ago [5], but has only recently gained widespread prominence as digital infrastructures became more integrated into industries, cities, and communities.

A Digital Twin, in its strictest sense, refers to a virtual counterpart of a physical process mirroring the real-time functioning of the physical entity [88]. This term was first introduced in the early 2000s by Michael Grieves [88], who used the concept in the production engineering field. However, since its introduction, the idea has evolved and broadened, now encompassing various forms of digital simulation models that are realigned through high-tech sensors to real-world processes, including those related to social and economic systems.

A Digital Twin, which replicates the real-time operations of a physical system, offers a perfect environment where RL agents can operate and learn, without the risks and constraints associated with experimenting in the physical world. Indeed, the synergy between Digital Twins and RL lies in the ability of the twin to serve as a virtual testing ground, allowing RL algorithms to explore various operational scenarios, test decisions, and improve over time without real-world consequences.

One of the key advantages of employing RL in Digital Twins is the ability to autonomously and adaptively manage complex systems. For instance, in industrial applications, RL can be applied to Digital Twins of manufacturing processes to optimize resource allocation, reduce downtime, and enhance production efficiency [172, 265]. Similarly, in smart cities, Digital Twins of infrastructure systems, such as transportation or energy grids, can be paired with RL algorithms to optimize traffic flows, reduce energy consumption, and respond to real-time demands or disruptions [208].

The practical implementation of RL and Digital Twins faces numerous challenges, among which the most critical is the "simulation-to-reality" gap, commonly referred to as the "reality gap" [4]. Even minor discrepancies in physical attributes, such as friction, noise levels, or sensor accuracy, can cause ML models, which performed successfully in simulated environments, to struggle in real-world contexts. Consequently, a meticulous approach is necessary when deploying RL agents trained in simulated settings to manage or optimize physical systems. While Digital

Twins are designed to replicate real-world systems with high precision, they cannot fully capture every specific characteristic of these systems. Nevertheless, the advantages of using Digital Twins to mitigate the reality gap have been demonstrated. Specifically, tailored training methodologies can enhance RL agents' performance in real-world applications, including autonomous driving [251] and robotic grasping [141].

Despite the potential of RL to achieve substantial outcomes when integrated with Digital Twins, it is essential to recognize that removing the human-in-the-loop role from the decision-making process may introduce significant complications. In high-risk scenarios, an RL agent might perform an erroneous action due to various unpredictable factors, potentially resulting in substantial losses. Therefore, in such contexts, RL can serve in a Decision Support System, offering possible solutions or policy recommendations for addressing complex challenges. However, the final decision should remain with the human decision-maker. Thus, the integration of RL and Digital Twins can serve a dual role: autonomously control specific processes and providing suggestions to the users in high risk situations.

Nonetheless, as already discussed in Chap. 3, the RL agents that can autonomously learn and optimize complex systems are inherently "black-boxes", and thus, the decision-making process is often opaque. This lack of interpretability/explainability can limit the adoption of RL for supporting the decision maker in high-risk applications, particularly in domains such as healthcare, finance, or critical infrastructure, where human operators need to understand and trust the actions recommended by RL systems. Among these applications, Digital Twins represent one of the main framework where this issue is remarkably relevant.

Therefore, the adoption of XRL methodologies in Digital Twins offers a pathway to greater transparency and insight into the behavior of RL agents operating first within a simulated environment and then in the physical system. Moreover, the training complications are avoided by the utilization of the virtual environment generated by the Digital Twin and the risk derived from exploiting a "black-box" agent to the real-world case scenarios is decreased through the XRL approaches. Additionally, a clear explanation of the action selection process can provide crucial information for the supporting the final decision that the human user has to take.

Following this idea, in the remaining part of the chapter, the Auto-BENEDICT method is applied to a specific scenario Digital Twin: the Robotic Arm environment.

## 7.2    A Case Study of Digital Twin: The Robotic Arm Environment

Before presenting the computational results attained by the Auto-BENEDICT method, it is necessary to introduce the Digital Twin considered: the Robotic Arm environment.

The task, that has been used for testing the methodologies developed in this thesis, concerns the movement of a package to a particular spot using a Six-axis Mirobot Robotic Arm. This problem simplifies the real-world scenarios where robotic arms have to move objects around without breaking them, e.g., automated warehouse management. Specifically, the RL agent has to control the gripper maintaining it elevated from the ground and moving it around a fixed space, without colliding with any obstacles. The main goal consists of picking up the package, which is in one of the corners of the map, and bringing it to the final location. Nevertheless, the agent's training can face multiple limitations if directly applied to the mechanical systems, e.g., the long time required for operating the actions and the possibility of damaging the robotic arm when bumping into the obstacles. For this reason, a Digital Twin of the physical mechanism has been created, allowing to simulate the training only for the virtual representation (shown in Fig. 7.1). Thus, a consistent reduction of the training time and of the risk of breaking some components of the machine is attained.

The digital representation of the robotic arm has been produced through the adoption of the WLKATA Studio software[1]. This specific software allows the user to receive the angle/coordinate of the physical system when plugged into the computer, and thus, control the robotic arm. Moreover, it includes a Python programming space where it can be directly instructed to solve particular tasks. In this way, the complete implementation of the environment has been facilitated. However, the complexity of the problem remains high due to the continuous variables involved.

For this reason, from the Digital Twin, a discrete version has been generated in order to further simplify the task to the RL agent. In detail, the environment consists of a $10 \times 10$ grid map with 5 different height levels, as shown in Fig. 7.1c. The episode starts with the gripper in the upper left corner of the map and at the 0 level of height, the package at the bottom left corner (on the ground), with the end-goal position at the bottom right corner. Multiple obstacles, spanning from the ground to the highest point, are positioned in the junctions of the map, leading to the creation of zones reachable only through bottleneck passages. In fact, it must be pointed out

---

[1] https://academy.wlkata.com/course/mirobot-programming-and-control

that the central area (in Fig. 7.1c) can not be used since it is occupied by the base
of the robotic arm.

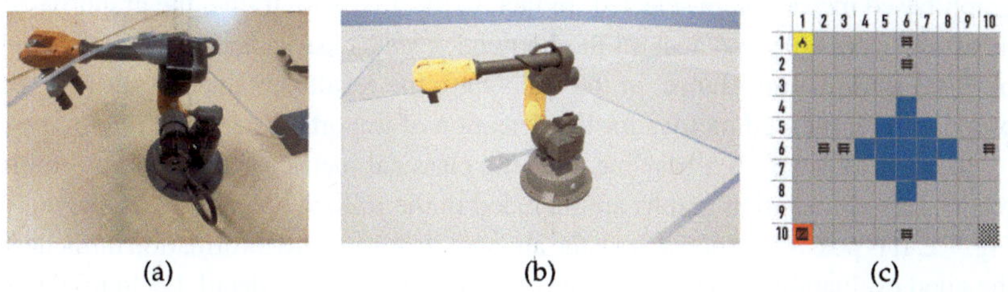

**Fig. 7.1** Robotic Arm Environment: (a) Physical system, (b) Digital Twin, (c) Grid-world
environment (at ground level)

Formally, the state space is composed of three-dimensional coordinates indicating the position of the robotic arm and whether the package has been picked up.
The possible actions chosen by the agent are a movement in each direction (up,
down, north, south, east, west) and a specific operation of picking up or dropping
off the package. The reward scheme is similar to the Taxi environment: for each
step, a reward of $-1$ is obtained, and only under specific circumstances the reward
is different. Explicitly, if the RL agent decides to pick up or drop off the package
in the wrong position, then it earns $-5$, while if the pick up or drop off operation
is successfully done, a reward of 5 and 20 respectively is obtained. Moreover, a
penalty of $-10$ is given to the agent if it collides against a wall of the map, or if it
bumps into an obstacle or the base of the robotic arm. Lastly, to avoid dragging the
package on the ground, a reward of $-3$ is given if the movements are operated at
the floor level (height 0) after picking up the object. The maximum number of steps
for each episode is set to 200 to cut off long and inefficient trials.

After introducing the characteristics of the Digital Twin environment used for
the computational evaluations, the results of the application of Auto-BENEDICT
are reported in the following section.

## 7.3   Results in the Robotic Arm Environment

For solving the Robotic Arm task, a Q-learning agent has been trained for 50,000
episodes. The trajectories, stored from the training phase, are used for the generation
of the main components of the Auto-BENEDICT approach.

The first step consists of defining the relevant states of the Robotic Arm environment. To this aim, an advanced analysis of the importance metrics has been performed. Given the similarities of this problem to the Taxi with traffic environment, graph-based metrics were selected. In fact, in the Taxi with traffic, the graph-based measures were able to detect all the bottleneck states associated with the concept of distal information. Hence, in the following, the results obtained evaluating the proposed graph-based metrics for the detection of important states are reported.

In order to establish a baseline, also the classical metrics applied to the simple reconstructed transition graphs are included in the following analysis, as shown in Fig. 7.2. It is possible to notice how the absence of any reward information influences the good evaluation of the importance of each state. More in detail, beginning with the betweenness centrality results presented in Fig. 7.2a, the major spots highlighted are the location of the package and the final destination. Nevertheless, the states that are recognized as the most important ones are always at the ground level. Moreover, it does not even change the perspective if the package has been already picked up or not.

A different outcome is generated for the closeness centrality shown in Fig. 7.2b where the importance is monotonically decreasing with the height of the grip. Additionally, the positions in the lower two sections of the map are recognized as highly relevant. However, due to the penalty parameter associated with dragging the package on the floor, the ground level should be less important than the height level 1 for all the positions different from the pick up and drop off location. Despite the aforementioned pitfall, the closeness centrality metric applied to the simple graph can identify the barriers and the base of the robotic arm as states to avoid. This is a direct consequence of terminating the episode in case of collision, inducing these dangerous nodes to not have any other connections, hence, showing a lower closeness centrality measure.

Moving to the results of the eigenvector centrality metric reported in Fig. 7.2c, it can be noticed that no insights are obtained from this computation, making it unsuitable for this scenario.

Visible improvements are achieved with the application of the classical centrality metrics to the Reconstructed Transition Multigraph (where for the closeness and eigenvector measures it is set $\varepsilon_C = 0.1$), as shown in Fig. 7.3.

In the case of the betweenness centrality in Fig. 7.3a, the addition of the reward structure of the MDP has brought the correct evaluation of the shortest path. Indeed, the optimal trajectory to follow after grabbing the object, i.e., elevating the grip at the first height level and then moving towards the destination is correctly acknowledged. Moreover, the salient spots of the maps (where to pick up the package and where to drop it off) are detected as important in all the different circumstances.

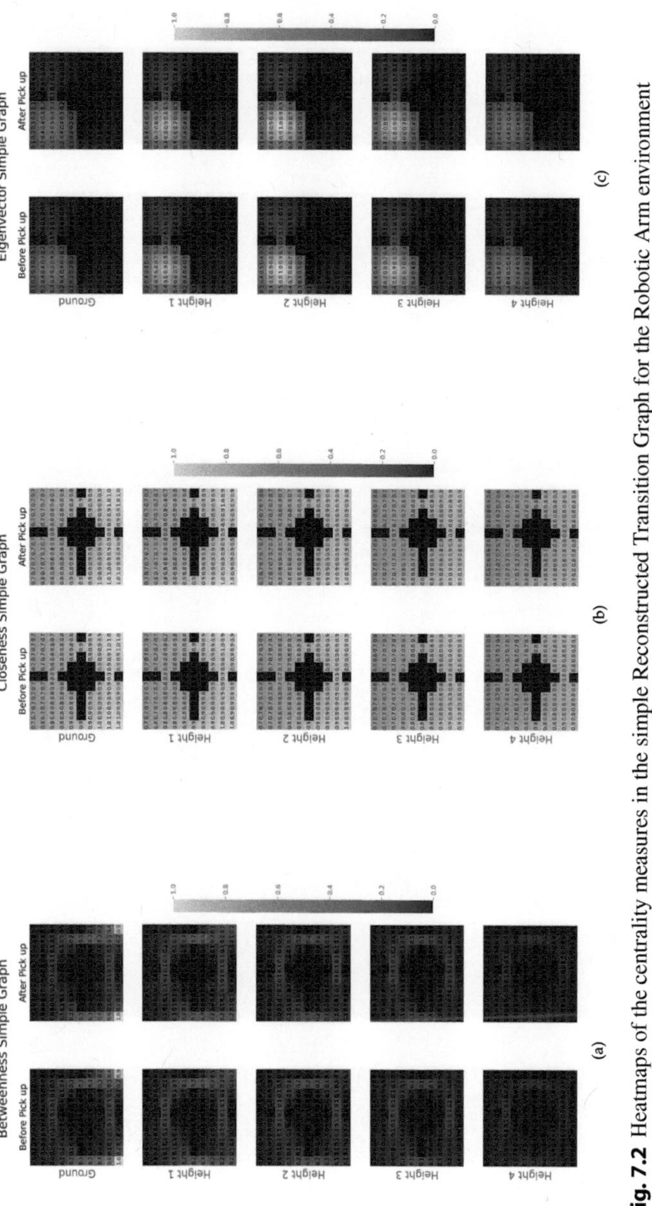

**Fig. 7.2** Heatmaps of the centrality measures in the simple Reconstructed Transition Graph for the Robotic Arm environment

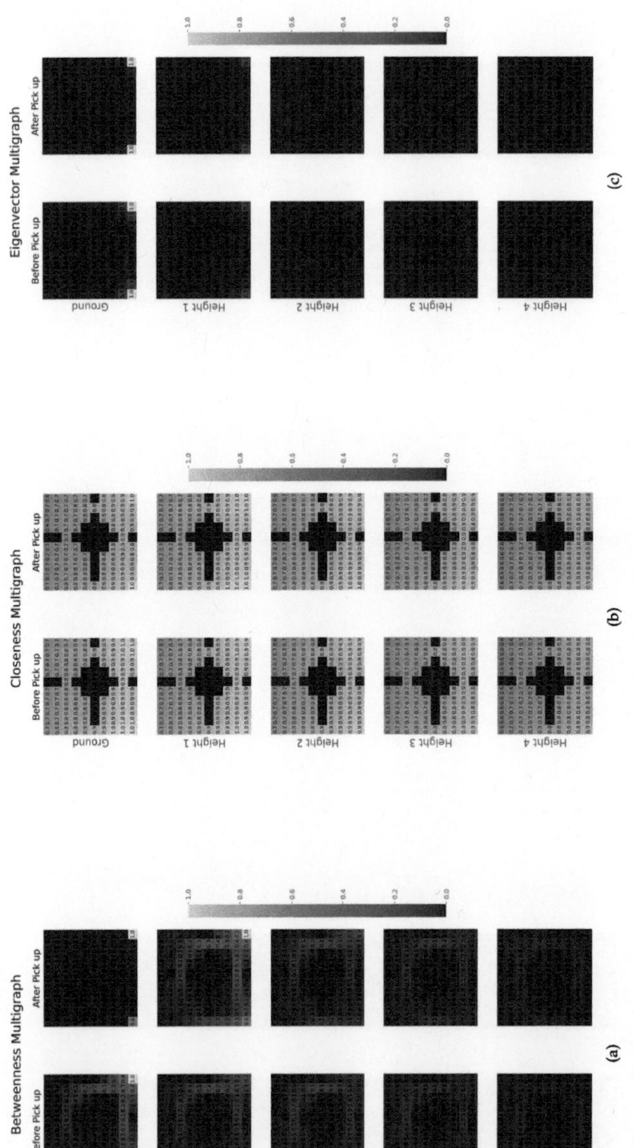


**Fig. 7.3** Heatmaps of the centrality measures in the Reconstructed Transition Multigraph for the Robotic Arm environment

Refinements on the evaluation of the importance of each state are achieved also by the closeness centrality, as displayed in Fig. 7.3b. Here, the major difference to the simple graph results consists of undervaluing the states that are at the ground level after picking up the package. The only positions that are relevant at the ground levels are the initial package location and the destination, since they are the starting and ending points of the optimal trajectory.

For the eigenvector centrality instead, not many improvements are observed in Fig. 7.3c. The only modification from the simple graph measures is the identification of the package location and the destination. Therefore, this measure can not attain satisfying results in the recognition of the fundamental states.

As for the computational experiments carried out in the Taxi environment with traffic (see Chap. 5), also in this context the classical centrality measures employed in the Reconstructed Transition Multiplex do not show exceptional performances. Here, the constant $\varepsilon_C$ is set to $0.1$, while the weight for the inter-layer connections $\varepsilon_I$ is set to $0.01$. In order to understand the quality of the evaluation, it is necessary to bear in mind that, for the multiplex measures, the importance measures are evaluating state-action pairs. Therefore, it is possible to discriminate the usefulness of a centrality measure by checking whether the high importance state-action pairs are belonging to the best trajectories.

In the case of the betweenness centrality (Fig. 7.4–7.5), the optimal trajectory to follow to complete the task can be noted. The salient moments of the MDP are also correctly identified: when to pick up and drop off the package. However, the problem of attaching an intuitive explanation of why the actions should be swapped with their counterparts still remains. This issue is present in all the classical centrality measures for the Reconstructed Transition Multiplex. Thus, this represents a fundamental question that must be addressed in the future.

For what concerns the closeness centrality results shown in Fig. 7.6–7.7, it can be noted the difference between the importance of the states at the ground level before and after picking up the package. Consequently, these measures still produce better insights than the ones applied to the simple graph. Moreover, it is noticeable that the pick up and drop off locations are more highlighted (with respect to the neighboring states) in these experiments than in the outcomes computed in the Reconstructed Transition Multigraph.

Similar conclusions can be traced for the versatility measure, reported in Fig. 7.8–7.9.

The only eigenvector metric that can attain decent performances in recognizing the critical states is the $f$-eigenvector metric, as displayed in Fig. 7.10. It is able to recognize the destination at the ground level as fundamental and all the remaining states at that height as dangerous. However, this measure gives higher importance

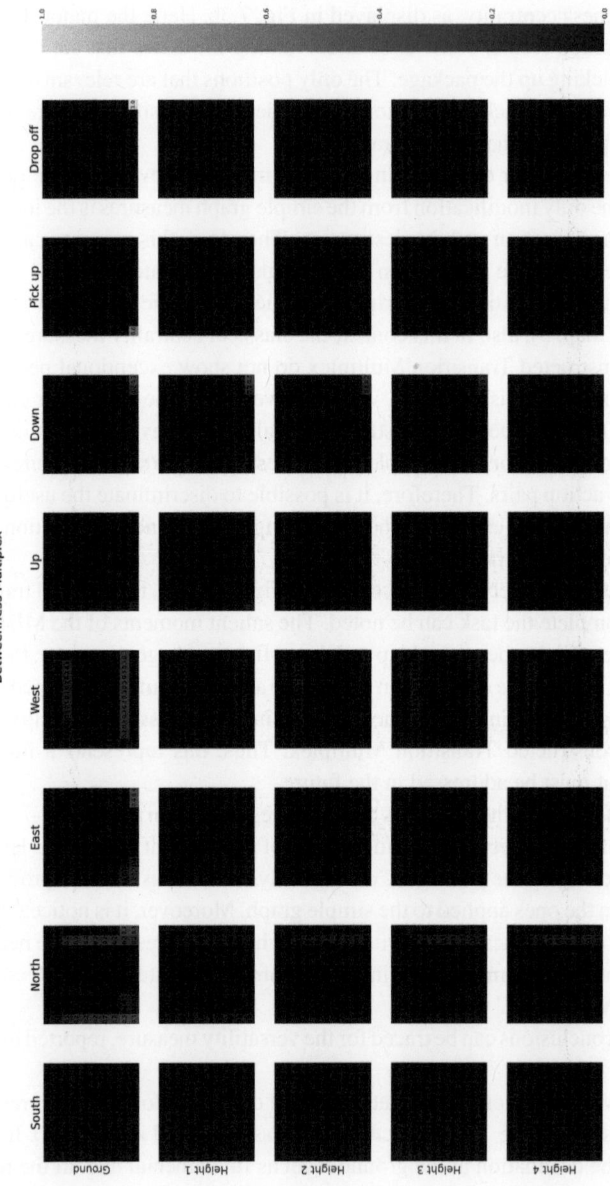

**Fig. 7.4** Heatmaps of the betweenness centrality in the Reconstructed Transition Multiplex for the Robotic Arm environment before picking up the package

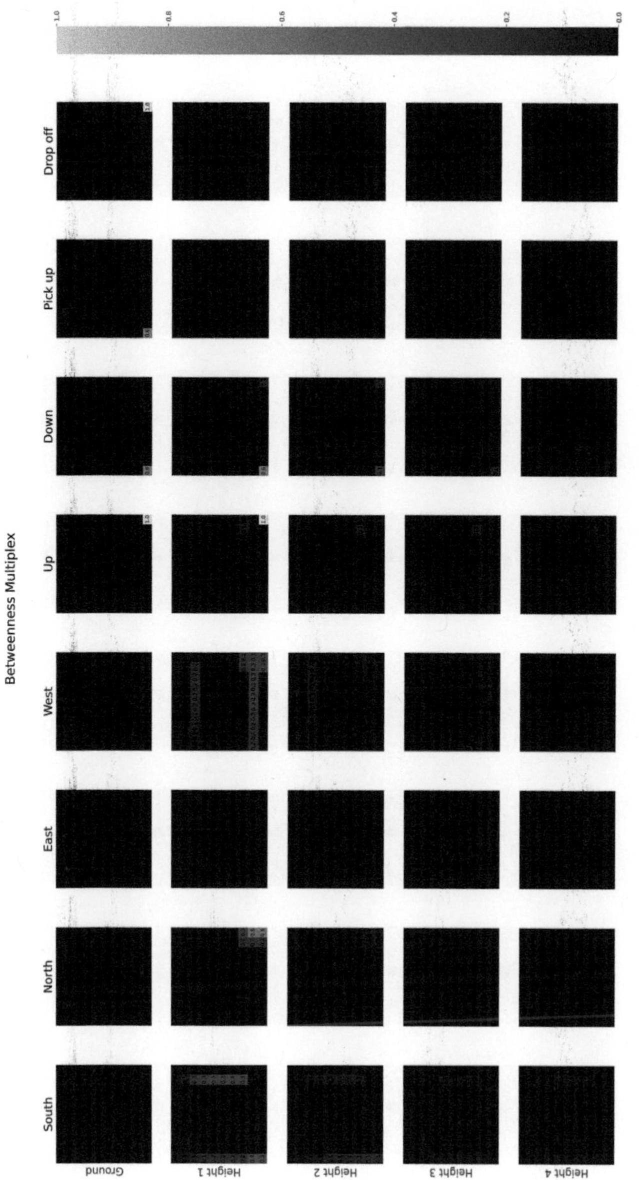


**Fig. 7.5** Heatmaps of the betweenness centrality in the Reconstructed Transition Multiplex for the Robotic Arm environment after picking up the package

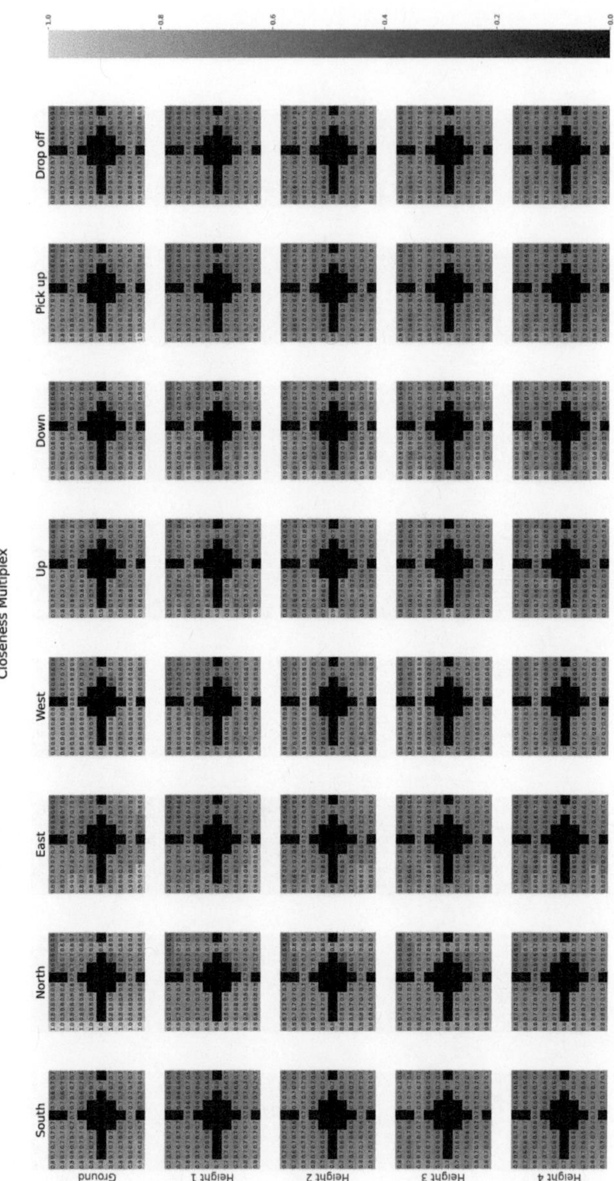



**Fig. 7.6** Heatmaps of the closeness centrality in the Reconstructed Transition Multiplex for the Robotic Arm environment before picking up the package

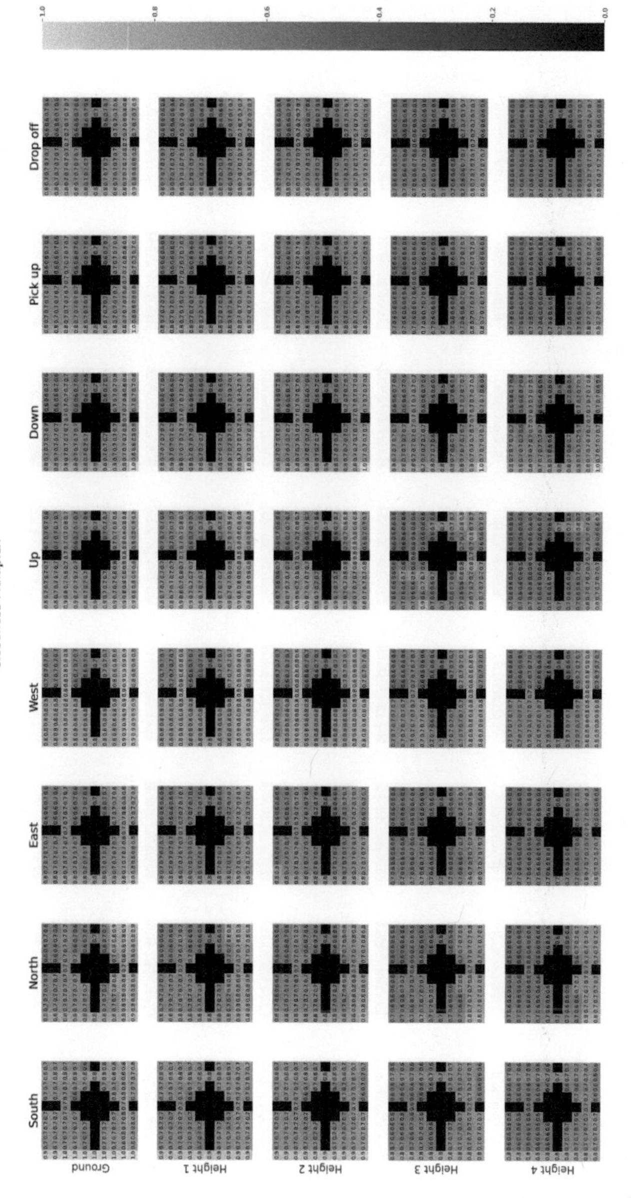

**Fig. 7.7** Heatmaps of the closeness centrality in the Reconstructed Transition Multiplex for the Robotic Arm environment after picking up the package

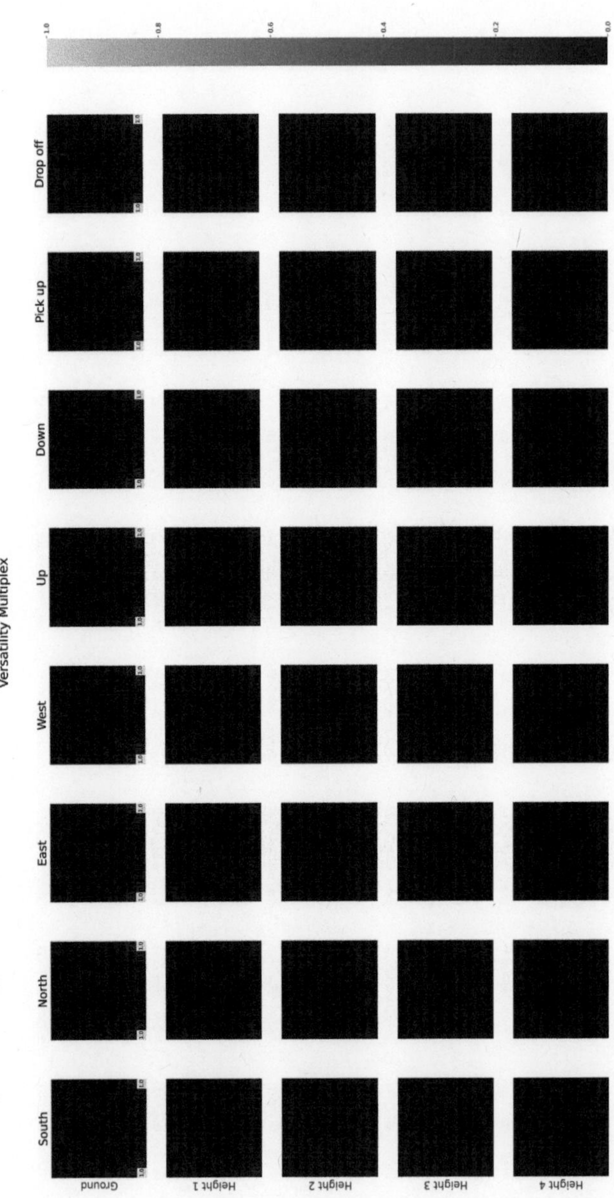


**Fig. 7.8** Heatmaps of the versatility measure in the Reconstructed Transition Multiplex for the Robotic Arm environment before picking up the package

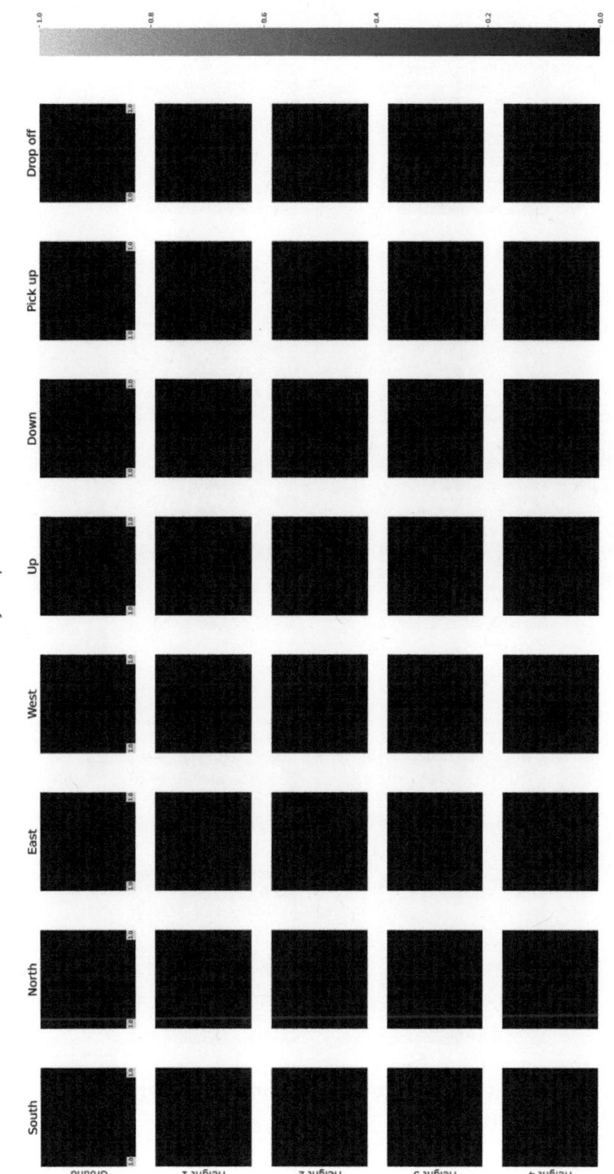

**Fig. 7.9** Heatmaps of the versatility measure in the Reconstructed Transition Multiplex for the Robotic Arm environment after picking up the package

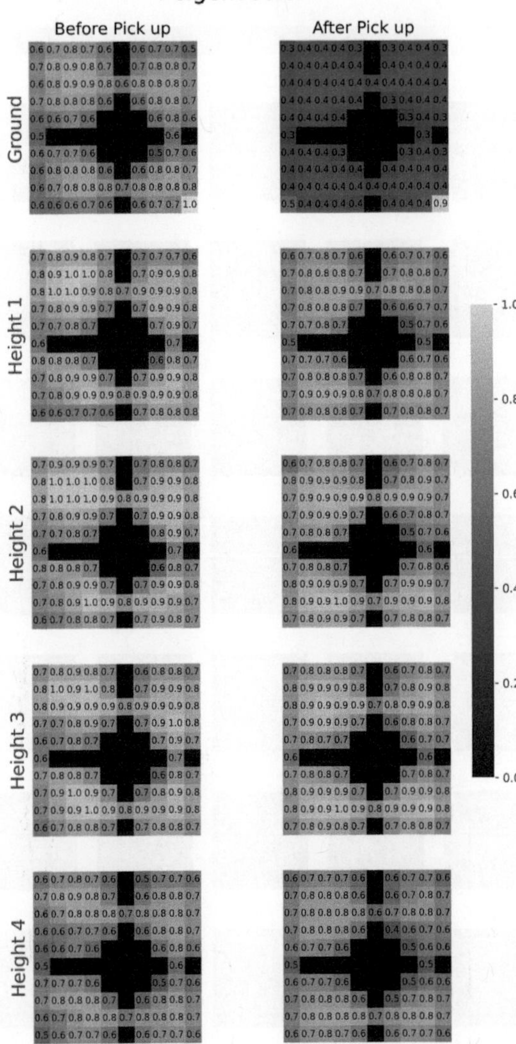

**Fig. 7.10** Heatmaps of the $f$-eigenvector centrality in the Reconstructed Transition Multiplex for the Robotic Arm environment

to the states that are well connected with others. For this reason, the positions at the second level of height present the largest importance values.

Given the results discussed above, the betweenness centrality for the Reconstructed Transition Multigraph has been selected for the discrimination of the relevant states. By applying the procedure described in Chap. 6, it has been possible to correctly determine the bottlenecks as the relevant state and use them as distal states.

Moreover, in order to define the risk levels, the $f$-eigenvector metric has been considered. In this case, a high value of the importance measure indicates a low level of risk. Hence, the level risk formulated for the Difference and the IQVA-D metrics of the previous chapter have to be reversed. Specifically to this experiment, a low level of risk is associated with the states presenting a value higher than the median ($> 0.67$), a medium level of risk to the states with a value between the median ($0.67$) and the first quartile ($0.40$), and, lastly, a high level of risk to the states with a value lower than the first quartile ($< 0.40$).

After delineating the information regarding the importance measures, it is possible to shift the focus to the creation of the BN for the predictions of future changes and the discovery of the causal relationships. The BN found applying the NOTEARS-MDP algorithm with $\varepsilon = 0.5$ is shown in Fig. 7.11. Also in this case, similitude can be observed between the DAG found for the Robotic Arm problem and the causal structure obtained from the same procedure in the Taxi environment (in Fig. 4.9).

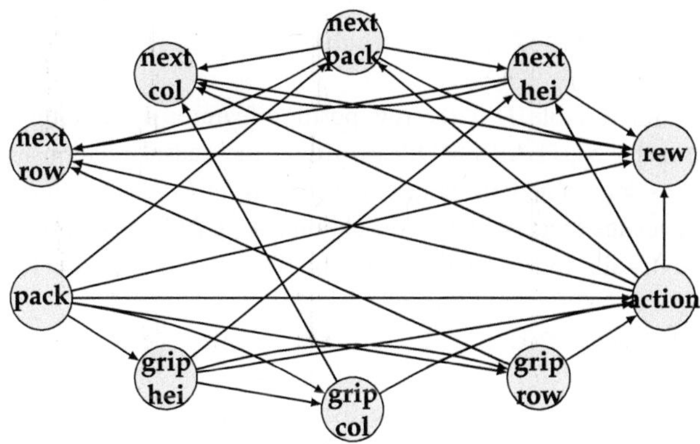


**Fig. 7.11** DAG obtained from NOTEARS-MDP with $\varepsilon = 0.5$ for the Robotic Arm environment

In the same way, as the destination plays a fundamental role in causally determining changes in the position of the taxi, in this scenario the package position has a central role in influencing the row, column and height position of the grip. Moreover, it can be noted how the height of the grip influences the row and column position features. This is explained by the penalty associated with moving the grip at the ground level after picking up the package. Therefore, just after picking the object, it is fundamental to increase the height of the grip.

Regarding the prediction accuracy of the BN, an average of 85,86% of correct predictions for all the future features and selected actions have been obtained. Specifically, this evaluation has been tested over 100 episodes, as usually done in the literature [148, 149].

After determining the multiple components of the Auto-BENEDICT procedure, it is now possible to show an example of explanation created in the Robotic Arm environment. The actual state instantiation analyzed, portrayed in Fig. 7.1c, is mathematically expressed as follows:

$$s_t = (\texttt{Row 1}, \texttt{Column 1}, \texttt{Height 1}, \texttt{at the Ground}) \tag{7.1}$$

where the grip is in the upper-left corner at the ground level, while the package is still at the ground level in its original position. The action chosen by the agent is $\texttt{Go South}$. Thus, the explanation has to answer the question "Why $\texttt{Go South}$?". Using the BN in Fig. 7.11, the next state features are predicted:

$$P_{act} = (\texttt{Row 2}, \texttt{Column 1}, \texttt{Height 1}, \texttt{at the Ground}). \tag{7.2}$$

Then, the central variable $D$ is the row position feature. By looking at the DAG obtained from the NOTEARS-MDP algorithm, the causal relationship with the other features is determined: the nodes in the parent set $Pa(D)$ are the height of the grip and the package location, while the children set $Ch(D)$ is empty.

The actual distal information forecasted by the actual RNN is the following:

$$d_{act} = \big((\texttt{Row 6}, \texttt{Column 1}, \texttt{Ground}, \texttt{at the Ground}), 1\big) \tag{7.3}$$

where the first piece of information is the distal state and the second value represents the final return attained by the agent.

The final information that need to be evaluated is the risk level. In this case, the $f$-eigenvector measure has a value of 0.70 for the actual instantiation $s_t$. Hence, the associated risk level is $\texttt{low level}$ since 0.70 > 0.67 (larger than the third quartile).

Finally, the textual answer is built following the pipeline of Sect. 4.3.4 and the final result is:

"Since the risk is at a `low level`, `Grip Height` is `Height 1` and `Package` is `at the Ground`, it is desirable to do action `Go South` in order to change the `Grip Row` from `Row 1` to `Row 2`, because `Package`, `Grip Height`, and `Grip row` are connected to the goal. Following this action, it will be possible to reach the state (`Row 6`, `Column 1`, `Ground`, `at the Ground`) and obtain a final return of 1".

Similarly, the counterfactual questions can be answered by including information regarding the counterfactual predictions of the BN and RNN.

## 7.4    Discussion

The application of the Auto-BENEDICT approach to the Robotic Arm environment demonstrates both the flexibility and effectiveness of combining causal explanation generation with importance metrics to support decision-making in Digital Twins. The methodology produces comprehensible, risk-aware explanations in a simulated industrial environment, enhancing transparency for users in critical decision scenarios. Thereby, the decision-makers are able to verify the action considered by the agent before performing it in the physical system.

In more detail, key metrics like the betweenness centrality in the Reconstructed Transition Multigraph and the $f$-eigenvector centrality in the Reconstructed Transition Multiplex frameworks allowed for effective identification of bottleneck states and assessment of associated risks. This critical insight into the structural importance of states directly contributed to making the generated explanations more actionable. By categorizing states into varying levels of risk, the users are provided with the tools to better understand the reasoning behind the RL agent's choices, particularly in handling hazardous decision points. This information, combined with the predictions of the BN and RNNs, produced risk-aware explanations reporting insights on the action selection process. This enhanced explanation capability is crucial for maintaining trust in the automation process, particularly as industries increasingly adopt Digital Twin technology in high-stakes environments like manufacturing and healthcare.

However, there is still room for improvement since the robustness of the distal information prediction remains constrained due to current limitations in prediction accuracy. This automation, while effective, suggests the need for further research into more efficient and precise training datasets that could better encompass the distal states forecasting without manual intervention. Furthermore, the measurements of the centrality metrics for the Reconstructed Transition Multiplex have to be improved in order to achieve acceptable results.

On the other hand, the $f$-eigenvector centrality of the Reconstructed Transition Multiplex achieves satisfying results in determining the risky states and also the highly connected states. Thereby, it can be considered for both evaluating the criticality of a state and the relevance of visiting the state for optimally solving the MDP. The combination of these properties is obtained only for the $f$-eigenvector centrality, indicating how the use of the Reconstructed Transition Multiplex can bring deeper insights in the tackled problem. Therefore, exploring other approaches considering the Reconstructed Transition Multiplex represents a fruitful future direction of research.

# Conclusion and Future Work 8

The central topic concerning this work was fully automatizing the creation of causal explanations for model-free RL agents. The resulting textual answers to "Why" and "Why not" questions had to maintain a high level of understandability in order to improve the trust of users in the algorithmic choices. To achieve this goal, a complex ensemble methodology has been developed: Auto-BENEDICT. The conceptualized pipeline consists of a combination of various methods coming from different fields: from the BNs for the causal reasoning to the RNNs for the distal predictions, passing through a wide set of importance metrics. The final approach shows comprehensible qualities for the outputted explanations and efficient forecasts for both the distal information and the next state instantiations. Moreover, the number of contributions related to preparing the Auto-BENEDICT method is extensive. A selection of the most important research developments and associated outcomes is presented in the following.

## 8.1    Contributions

The first contribution that has been encountered in this thesis is represented by the definition of a new classification scheme for the XRL methodologies. So far, it is not possible to clearly state a taxonomy of the approaches present in the literature due to the intricacy of the algorithms used. Hence, it has been suggested a division into classes that consider the different methods and outputs generated. Moreover, it has been analyzed thoroughly what are the salient characteristics that have been utilized in the state-of-the-art research to recognize the direction of the XRL field. By doing so, issues related to the automatization and human-validation attributes have been pointed out.


R. Milani, *Advanced Automation for Comprehensible Causal Explanations of Reinforcement Learning Agents*,
https://doi.org/10.1007/978-3-658-50495-3_8

Moving to the first practical result, the NOTEARS algorithm useful for the determination of DAGs has been improved including a pruning phase during its thresholding procedure. The pruning rules applied have been selected from the characteristics of the MDPs. Thereby, the graph generated from the data can approximate better the problems analyzed and satisfy the specific conditions of the MDPs. With this particular improvement, it has been possible to create BNs able to outperform both classical ML algorithms and specialized forecasting approaches in the prediction tasks of two out of three environments. Additionally, a novel procedure involving BNs for creating causal explanations for both actual and counterfactual scenarios is theorized.

Furthermore, the distal information used in the explanation has been enlarged in order to include also the distal states and the distal values, e.g., the final return. The datasets created for the predictions of these data for both the "Why" and "Why not" questions were modified from the ones suggested in the literature. In detail, additional variables have been included to facilitate the forecasting of the final return: for the value-based algorithm, the Q-values at the actual instantiation were used, while for the policy-based methods, the accumulated rewards until that time step has been exploited. In both cases, satisfying results, in terms of accuracy of the predictions for the distal actions or states, and mean square error for the final return, have been achieved. In this way, the problem of applying the proposed methodology to an RL agent that did not calculate the Q-values has been solved.

The explanations produced by the combinations of the improvements described above were also tested in a human validation study showing an increase in understandability and trust. In fact, for all the seven qualities that have been checked, the suggested statements outperformed the answers to "Why" questions created by methods from the literature. On the other hand, in the evaluation for the counterfactual explanations, only in three out of seven attributes, it has been proved a statistically significant difference between the output of the BENEDICT approach and the action influence models. Nonetheless, the majority of the tested users marked as favourite the proposed explanation, recognizing it as the one presenting enough information and being comprehensible.

For the discovery of the distal states, new importance metrics have been crafted. Both Q-value-based and graph-based methods have been improved. In this way, all the algorithms in the model-free group can be covered. Specifically for the former type, it has been introduced a new metric derived directly from the combination of the already existing measures: the difference metric. Through these freshly developed values, it is possible to discriminate high-risk states and increase the focus of the user on the action that has to be performed. Moreover, the classic metrics are modified to incorporate the frequency of visitation of each state and the time, when it has been encountered, in the final importance values. To accomplish this task, an iterated version of these measures is derived. With this particular implementation, the approach IQVA led to the discovery of bottleneck states fundamental for the completion of the MDPs.

On the other side, for the graph-based measurements, the concept of reconstructed transition graphs has been adapted to more complex structures: the multigraphs and multiplexes. Thereby, the graphs obtained are able to incorporate the reward description and evaluate more consistently the importance of each state or state-action. In this scenario, the theoretical guarantees for the existence and uniqueness of the associated centrality measures considered (betweenness, closeness and eigenvector) have been proved. The computational experiments show high efficiency, especially for the $f$−eigenvector centrality applied to the Reconstructed Transition Multiplexes. This adapted third-order metric is indeed able to discriminate highly connected nodes, bottlenecks and also dangerous states; showing extraordinary potential in the determination of relevant data.

The novel measures discussed above are then applied to the task of automatically discovery the distal information for the creation of the RNNs training datasets. Furthermore, the difference metric and its iterated version have been employed for the evaluation of the risk level for the actual state. Through this gimmick, the final explanations, outputted by the Auto-BENEDICT approach, can include an estimation of the danger at the actual time step, showing to the user that the next choice has to be decided carefully.

## 8.2   Future Works

In the remaining part of this chapter, interesting future directions of research are outlined.

Following the structure of this thesis, it can be outlined a series of grey areas that can be explored. For example, it can be useful to conceptualize a statistical methodology that can learn DAGs for causal graphs in the context of MDPs, without having to correct the results at a later stage. In particular, the properties that have been utilized in the pruning phase can be translated into specific constraints of the combinatorial optimization program used in the NOTEARS algorithm.

Remaining in the same field, a different improvement for the causal explanations could derive from the inclusion of the BN's probabilities of the prediction of the next state. In this way, it can be approximated the confidence of the BN in forecasting both the agent actions and the environmental responses. Obviously, this addition has to be evaluated in a human study, to provide statistically significant results.

As well as in the process of generation of the explanations, also in the definition of the importance metrics there is still room for improvement. A particularly interesting idea concerns the combination of the Q-value-based and graph-based measurements in order to attain good evaluations of both environmental conditions and the agent's knowledge. As testified by the correlation analysis executed, the two groups of metrics do not approximate the same ideological qualities, leading to requiring multiple of them to have a complete overview of the circumstance. Nevertheless, given the flexibility of the Reconstructed Transition Multigraphs and Multiplexes, it is possible to incorporate in the edges' weights information regarding the Q-values. Thus, this peculiar solution could try to bridge between the two classes, to reduce the gap of information.

A last future research direction connected to the results presented in this thesis considers the human-validation of the finally developed explanations. Indeed, this process has been avoided due to the following observation: the general structure of the textual output barely changes from the positively remarked explanations created with the BENEDICT approach. Hence, it can be assumed that similar results can be obtained if the last explanation proposed has been compared to the one in the literature.

A significant limitation in the field of XRL is the absence of a standardized benchmark for evaluating the quality and effectiveness of explanations. As seen throughout this thesis, various environments were employed to validate the proposed methodologies. However, this gap is prevalent across the literature, where a broad spectrum of problem domains, from simple control tasks and games to complex healthcare data, are used for evaluation. This diversity presents a challenge in comparing explanations and corresponding results, particularly for context-sensitive approaches. Therefore, establishing standardized environments and metrics for validating XRL methods is a critical objective that could substantially advance the field.

Another promising direction involves integrating Large Language Models into the explanation-generation process for RL agents. Given the recent advancements and transformative impact of Large Language Models, their application in XRL holds potential for enhancing the clarity and expressiveness of generated explanations. However, the content of these explanations should remain grounded in the insights derived through XRL techniques, with Large Language Models primarily serving to improve the clarity and coherence of the articulated explanations.

# Bibliography


1. Amina Adadi and Mohammed Berrada. "Peeking Inside the Black-Box: A Survey on Explainable Artificial Intelligence (XAI)." In: *IEEE Access* 6 (Sept. 2018), pp. 52138–52160. ISSN: 21693536. https://doi.org/10.1109/ACCESS.2018.2870052.
2. Mamoun Alazab, Latif U Khan, Srinivas Koppu, Swarna Priya Ramu, M Iyapparaja, Parimala Boobalan, Thar Baker, Praveen Kumar Reddy Maddikunta, Thippa Reddy Gadekallu, and Ahamed Aljuhani. "Digital twins for healthcare 4.0—Recent advances, architecture, and open challenges." In: *IEEE Consumer Electronics Magazine* 12.6 (2022), pp. 29–37.
3. Alnour Alharin, Thanh Nam Doan, and Mina Sartipi. "Reinforcement learning interpretation methods: A survey." In: *IEEE Access* 8 (2020), pp. 171058–171077. ISSN: 21693536. https://doi.org/10.1109/ACCESS.2020.3023394.
4. Jean Pierre Allamaa, Panagiotis Patrinos, Herman Van der Auweraer, and Tong Duy Son. "Sim2real for autonomous vehicle control using executable digital twin." In: *IFAC-PapersOnLine* 55.24 (2022), pp. 385–391.
5. B Danette Allen. "Digital twins and living models at NASA." In: *Digital Twin Summit.* 2021.
6. Duane F Alwin and Robert M Hauser. "The decomposition of effects in path analysis." In: American sociological review (1975), pp. 37–47.
7. Dan Amir and Ofra Amir. "Highlights: Summarizing agent behavior to people." In: *Proceedings of the 17th International Conference on Autonomous Agents and MultiAgent Systems.* 2018, pp. 1168–1176.
8. Ofra Amir, Finale Doshi-Velez, and David Sarne. "Summarizing agent strategies." In: *Autonomous Agents and Multi-Agent Systems* 33.5 (2019), pp. 628–644.
9. Yotam Amitai, Guy Avni, and Ofra Amir. "ASQ-IT: Interactive explanations for reinforcement-learning agents." In: *arXiv preprint* arXiv:2301.09941 (2023).
10. Marcin Andrychowicz, Filip Wolski, Alex Ray, Jonas Schneider, Rachel Fong, Peter Welinder, Bob McGrew, Josh Tobin, OpenAI Pieter Abbeel, and Wojciech Zaremba. "Hindsight experience replay." In: *Advances in neural information processing systems* 30 (2017).
11. Plamen Angelov and Ronald Yager. "Simplified fuzzy rule-based systems using non-parametric antecedents and relative data density." In: *2011 IEEE Workshop on Evolving and Adaptive Intelligent Systems (EAIS).* IEEE. 2011, pp. 62–69.

12. Jac M Anthonisse. "The rush in a directed graph." In: *Stichting Mathematisch Centrum. Mathematische Besliskunde* BN 9/71 (1971).

13. Alejandro Barredo Arrieta, Natalia Díaz-Rodríguez, Javier Del Ser, Adrien Bennetot, Siham Tabik, Alberto Barbado, Salvador García, Sergio Gil-López, Daniel Molina, Richard Benjamins, et al. "Explainable Artificial Intelligence (XAI): Concepts, taxonomies, opportunities and challenges toward responsible AI." In: *Information fusion* 58 (2020), pp. 82–115.

14. Francis R Bach and Michael I Jordan. "Kernel independent component analysis." In: *Journal of machine learning research* 3.Jul (2002), pp. 1–48.

15. Sebastian Bach, Alexander Binder, Grégoire Montavon, Frederick Klauschen, Klaus-Robert Müller, and Wojciech Samek. "On pixel-wise explanations for non-linear classifier decisions by layer-wise relevance propagation." In: *PloS one* 10.7 (2015), e0130140.

16. Michael Bain and Claude Sammut. "A Framework for Behavioural Cloning." In: *Machine Intelligence 15*. 1995, pp. 103–129.

17. Ravindra B Bapat. "A max version of the Perron-Frobenius theorem." In: *Linear Algebra and its Applications* 275 (1998), pp. 3–18.

18. Imad A Basheer and Maha Hajmeer. "Artificial neural networks: fundamentals, computing, design, and application." In: *Journal of microbiological methods* 43.1 (2000), pp. 3–31.

19. Osbert Bastani, Jeevana Priya Inala, and Armando Solar-Lezama. "Interpretable, verifiable, and robust reinforcement learning via program synthesis." In: *International Workshop on Extending Explainable AI Beyond Deep Models and Classifiers*. Springer. 2020, pp. 207–228.

20. Michael Batty. *Digital twins*. 2018.

21. Alex Bavelas. "A mathematical model for group structures." In: *Human organization* 7.3 (1948), pp. 16–30.

22. Alex Bavelas. "Communication patterns in task-oriented groups." In: *The journal of the acoustical society of America* 22.6 (1950), pp. 725–730.

23. Murray A Beauchamp. "An improved index of centrality." In: Behavioral science 10.2 (1965), pp. 161–163.

24. Marc G Bellemare, Georg Ostrovski, Arthur Guez, Philip Thomas, and Rémi Munos. "Increasing the action gap: New operators for reinforcement learning." In: *Proceedings of the AAAI Conference on Artificial Intelligence*. Vol. 30. 1. 2016.

25. Richard E. Bellman. *Adaptive Control Processes*. Princeton: Princeton University Press, 1961. ISBN: 9781400874668. https://doi.org/10.1515/9781400874668. URL: https://doi.org/10.1515/9781400874668.

26. Richard Bellman. "A Markovian decision process." In: *Journal of mathematics and mechanics* (1957), pp. 679–684.

27. Michele Berlingerio, Michele Coscia, Fosca Giannotti, Anna Monreale, and Dino Pedreschi. "Foundations of multidimensional network analysis." In: *2011 international conference on advances in social networks analysis and mining*. IEEE. 2011, pp. 485–489.

28. Abraham Berman and Robert J Plemmons. *Nonnegative matrices in the mathematical sciences*. SIAM, 1994.

29. Benjamin Beyret, Ali Shafti, and A Aldo Faisal. "Dot-to-dot: Explainable hierarchical reinforcement learning for robotic manipulation." In: *2019 IEEE/RSJ International Conference on Intelligent Robots and Systems (IROS)*. IEEE. 2019, pp. 5014–5019.

30. Vincent D Blondel, Jean-Loup Guillaume, Renaud Lambiotte, and Etienne Lefebvre. "Fast unfolding of communities in large networks." In: *Journal of statistical mechanics: theory and experiment* 2008.10 (2008), P10008.

31. Stefano Boccaletti, Ginestra Bianconi, Regino Criado, Charo I Del Genio, Jesús Gómez-Gardenes, Miguel Romance, Irene Sendina-Nadal, Zhen Wang, and Massimiliano Zanin. "The structure and dynamics of multilayer networks." In: *Physics reports* 544.1 (2014), pp. 1–122.

32. Béla Bollobás. Modern graph theory. Vol. 184. Springer Science & Business Media, 1998.

33. Phillip Bonacich. "Factoring and weighting approaches to status scores and clique identification." In: *Journal of mathematical sociology* 2.1 (1972), pp. 113–120.

34. Carlo Bonferroni. "Teoria statistica delle classi e calcolo delle probabilita." In: *Pubblicazioni del R Istituto Superiore di Scienze Economiche e Commerciali di Firenze* 8 (1936), pp. 3–62.

35. MIKE Boyle. "Notes on the perron-frobenius theory of nonnegative matrices." In: *Dept. of Mathematics, University of Maryland, College Park, MD, USA, Web site:* http://www.math.umd.edu/~mboyle/courses/475sp05/spec.pdf (2015).

36. Andrew P Bradley. "The use of the area under the ROC curve in the evaluation of machine learning algorithms." In: *Pattern recognition* 30.7 (1997), pp. 1145–1159.

37. Nadia Burkart and Marco F Huber. *A Survey on the Explainability of Supervised Machine Learning.* 2021, pp. 245–317.

38. Béatrice Cahour and Jean-François Forzy. "Does projection into use improve trust and exploration? An example with a cruise control system." In: *Safety science* 47.9 (2009), pp. 1260–1270.

39. Eren Cakmak, Manuel Plank, Daniel S Calovi, Alex Jordan, and Daniel Keim. "Spatiotemporal clustering benchmark for collective animal behavior." In: *Proceedings of the 1st ACM SIGSPATIAL International Workshop on Animal Movement Ecology and Human Mobility.* 2021, pp. 5–8.

40. Arthur Charpentier, Romuald Elie, and Carl Remlinger. "Reinforcement learning in economics and finance." In: *Computational Economics* (2021), pp. 1–38.

41. Probal Chaudhuri, Min-Ching Huang, Wei-Yin Loh, and Ruji Yao. "Piecewise-polynomial regression trees." In: *Statistica Sinica* (1994), pp. 143–167.

42. Jessie YC Chen, Shan G Lakhmani, Kimberly Stowers, Anthony R Selkowitz, Julia L Wright, and Michael Barnes. "Situation awareness-based agent transparency and human-autonomy teaming effectiveness." In: *Theoretical issues in ergonomics science* 19.3 (2018), pp. 259–282.

43. Ziheng Chen, Fabrizio Silvestri, Gabriele Tolomei, Jia Wang, He Zhu, and Hongshik Ahn. "Explain the explainer: Interpreting model-agnostic counterfactual explanations of a deep reinforcement learning agent." In: *IEEE Transactions on Artificial Intelligence* (2022).

44. Zelei Cheng, Xian Wu, Jiahao Yu, Wenhai Sun, Wenbo Guo, and Xinyu Xing. "Statemask: Explaining deep reinforcement learning through state mask." In: *Thirty-seventh Conference on Neural Information Processing Systems.* 2023.

45. David Maxwell Chickering. "Learning Bayesian networks is NP-complete." In: *Learning from data.* Springer, 1996, pp. 121–130.

46. Junyoung Chung, Caglar Gulcehre, KyungHyun Cho, and Yoshua Bengio. "Empirical evaluation of gated recurrent neural networks on sequence modeling." In: *arXiv preprint* arXiv:1412.3555 (2014).

47. Jeffery Allen Clouse. *On integrating apprentice learning and reinforcement learning*. University of Massachusetts Amherst, 1996.

48. Youri Coppens, Kyriakos Efthymiadis, Tom Lenaerts, Ann Nowé, Tim Miller, Rosina Weber, and Daniele Magazzeni. "Distilling deep reinforcement learning policies in soft decision trees." In: *Proceedings of the IJCAI 2019 workshop on explainable artificial intelligence*. 2019, pp. 1–6.

49. Michele Coscia, Giulio Rossetti, Diego Pennacchioli, Damiano Ceccarelli, and Fosca Giannotti. ""You know because I know" a multidimensional network approach to human resources problem." In: *Proceedings of the 2013 IEEE/ACM International Conference on Advances in Social Networks Analysis and Mining*. 2013, pp. 434–441.

50. Francisco Cruz, Charlotte Young, Richard Dazeley, and Peter Vamplew. "Evaluating Human-like Explanations for Robot Actions in Reinforcement Learning Scenarios." In: *arXiv preprint* arXiv:2207.03214 (2022).

51. Leonardo L Custode and Giovanni Iacca. "Evolutionary learning of interpretable decision trees." In: *IEEE Access* 11 (2023), pp. 6169–6184.

52. George Cybenko. "Approximation by superpositions of a sigmoidal function." In: *Mathematics of control, signals and systems* 2.4 (1989), pp. 303–314.

53. Giang Dao, Indrajeet Mishra, and Minwoo Lee. "Deep reinforcement learning monitor for snapshot recording." In: *2018 17th IEEE International Conference on Machine Learning and Applications (ICMLA)*. IEEE. 2018, pp. 591–598.

54. Arun Das and Paul Rad. "Opportunities and challenges in explainable artificial intelligence (xai): A survey." In: *arXiv preprint* arXiv:2006.11371 (2020).

55. Sanjeeb Dash, Oktay Gunluk, and Dennis Wei. "Boolean decision rules via column generation." In: *Advances in neural information processing systems* 31 (2018).

56. Omid Davoodi and Majid Komeili. "Feature-Based Interpretable Reinforcement Learning based on State-Transition Models." In: *2021 IEEE International Conference on Systems, Man, and Cybernetics (SMC)*. IEEE. 2021, pp. 301–308.

57. Manlio De Domenico, Albert Solé-Ribalta, Elisa Omodei, Sergio Gómez, and Alex Arenas. "Centrality in interconnected multilayer networks." In: *arXiv preprint* arXiv:1311.2906 (2013).

58. Manlio De Domenico, Albert Solé-Ribalta, Elisa Omodei, Sergio Gómez, and Alex Arenas. "Ranking in interconnected multilayer networks reveals versatile nodes." In: *Nature communications* 6.1 (2015), p. 6868.

59. Thomas Dean and Keiji Kanazawa. "A model for reasoning about persistence and causation." In: *Computational intelligence* 5.2 (1989), pp. 142–150.

60. Quentin Delfosse, Hikaru Shindo, Devendra Dhami, and Kristian Kersting. "Interpretable and explainable logical policies via neurally guided symbolic abstraction." In: *Advances in Neural Information Processing Systems* 36 (2024).

61. Thomas G Dietterich. "Hierarchical reinforcement learning with the MAXQ value function decomposition." In: *Journal of artificial intelligence research* 13 (2000), pp. 227–303.

62. Bruce Digney. "Learning hierarchical control structures for multiple tasks and changing environments." In: *Proceedings of the fifth conference on the simulation of adaptive behavior: SAB*. Vol. 98. Citeseer. 1998, p. 295.

63. Jianguo Ding and A Rebai. "Probabilistic inferences in Bayesian networks." In: *Bayesian Network* (2010), pp. 39–53.

64. Finale Doshi-Velez and Been Kim. "Towards a rigorous science of interpretable machine learning." In: *arXiv preprint* arXiv:1702.08608 (2017).

65. Filip Karlo Došilović, Mario Brčić, and Nikica Hlupić. "Explainable artificial intelligence: A survey." In: *2018 41st International convention on information and communication technology, electronics and microelectronics (MIPRO)*. IEEE. 2018, pp. 0210–0215.

66. Sandrine Dudoit, Mark J Van Der Laan, and Mark J van der Laan. *Multiple testing procedures with applications to genomics*. Springer, 2008.

67. Gabriel Dulac-Arnold, Nir Levine, Daniel J Mankowitz, Jerry Li, Cosmin Paduraru, Sven Gowal, and Todd Hester. "Challenges of real-world reinforcement learning: definitions, benchmarks and analysis." In: *Machine Learning* 110.9 (2021), pp. 2419–2468.

68. Olive Jean Dunn. "Multiple comparisons among means." In: *Journal of the American statistical association* 56.293 (1961), pp. 52–64.

69. Martin Erwig, Alan Fern, Magesh Murali, and Anurag Koul. "Explaining deep adaptive programs via reward decomposition." In: *IJCAI/ECAI workshop on explainable artificial intelligence*. 2018.

70. Joshua B Evans and Özgür Şimşek. "Creating Multi-Level Skill Hierarchies in Reinforcement Learning." In: *arXiv preprint* arXiv:2306.09980 (2023).

71. Eugene A Feinberg and Adam Shwartz. *Handbook of Markov decision processes: methods and applications*. Vol. 40. Springer Science & Business Media, 2012.

72. Andrea Ferigo, Leonardo Lucio Custode, and Giovanni Iacca. "Quality–diversity optimization of decision trees for interpretable reinforcement learning." In: *Neural Computing and Applications* (2023), pp. 1–12.

73. Matthew C Fontaine, Julian Togelius, Stefanos Nikolaidis, and Amy K Hoover. "Covariance matrix adaptation for the rapid illumination of behavior space." In: *Proceedings of the 2020 genetic and evolutionary computation conference*. 2020, pp. 94–102.

74. Michel Fortin and Roland Glowinski. *Augmented Lagrangian methods: applications to the numerical solution of boundary-value problems*. Elsevier, 2000.

75. Linton C Freeman. "A set of measures of centrality based on betweenness." In: *Sociometry* (1977), pp. 35–41.

76. Alex A Freitas. "Comprehensible classification models: a position paper." In: *ACM SIGKDD explorations newsletter* 15.1 (2014), pp. 1–10.

77. Georg Frobenius, Ferdinand Georg Frobenius, Ferdinand Georg Frobenius, Ferdinand Georg Frobenius, and Germany Mathematician. *Über Matrizen aus nicht negativen Elementen*. Königliche Akademie der Wissenschaften Berlin, 1912.

78. Nicholas Frosst and Geoffrey Hinton. "Distilling a neural network into a soft decision tree." In: *arXiv preprint* arXiv:1711.09784 (2017).

79. Jasmina Gajcin and Ivana Dusparic. "ReCCoVER: Detecting Causal Confusion for Explainable Reinforcement Learning." In: *arXiv preprint* arXiv:2203.11211 (2022).

80. Everette S Gardner Jr. "Exponential smoothing: The state of the art." In: *Journal of forecasting* 4.1 (1985), pp. 1–28.

81. Hector Geffner, Rina Dechter, and Joseph Y Halpern. "Probabilistic and Causal Inference: The Works of Judea Pearl." In: ACM, 2022.

82. Alan Gibbons. *Algorithmic graph theory*. Cambridge university press, 1985.

83. Claire Glanois, Paul Weng, Matthieu Zimmer, Dong Li, Tianpei Yang, Jianye Hao, and Wulong Liu. "A Survey on Interpretable Reinforcement Learning." In: *arXiv preprint* arXiv:2112.13112 (2021).

84. Mariel K Goddu and Alison Gopnik. "The development of human causal learning and reasoning." In: *Nature Reviews Psychology* (2024), pp. 1–21.

85. Arthur S Goldberger. "Structural equation methods in the social sciences." In: *Econometrica: Journal of the Econometric Society* (1972), pp. 979–1001.

86. Megan Goldman. "Statistics for bioinformatics." In: *Pdf.[(accessed on 8 May 2024)]* (2008).

87. Samuel Greydanus, Anurag Koul, Jonathan Dodge, and Alan Fern. "Visualizing and understanding atari agents." In: *International conference on machine learning*. PMLR. 2018, pp. 1792–1801.

88. Michael Grieves. "Digital twin: manufacturing excellence through virtual factory replication." In: *White paper* 1.2014 (2014), pp. 1–7.

89. Riccardo Guidotti, Anna Monreale, Salvatore Ruggieri, Franco Turini, Fosca Giannotti, and Dino Pedreschi. "A survey of methods for explaining black box models." In: *ACM Computing Surveys* 51 (5 Aug. 2018). ISSN: 15577341. https://doi.org/10.1145/3236009.

90. David Gunning. "Explainable artificial intelligence (xai)." In: *Defense advanced research projects agency (DARPA), nd Web* 2.2 (2017), p. 1.

91. Wenbo Guo, Xian Wu, Usmann Khan, and Xinyu Xing. "Edge: Explaining deep reinforcement learning policies." In: *Advances in Neural Information Processing Systems* 34 (2021), pp. 12222–12236.

92. Piyush Gupta, Nikaash Puri, Sukriti Verma, Sameer Singh, Dhruv Kayastha, Shripad Deshmukh, and Balaji Krishnamurthy. "Explain your move: Understanding agent actions using focused feature saliency." In: *arXiv preprint* arXiv:1912.12191 (2019).

93. David Ha and Jürgen Schmidhuber. "Recurrent world models facilitate policy evolution." In: *Advances in neural information processing systems* 31 (2018).

94. Ammar Haydari and Yasin Yilmaz. "Deep Reinforcement Learning for Intelligent Transportation Systems: A Survey." In: *IEEE Transactions on Intelligent Transportation Systems* 23 (1 Jan. 2022), pp. 11–32. ISSN: 15580016. https://doi.org/10.1109/TITS.2020.3008612.

95. Bradley Hayes and Julie A Shah. "Improving robot controller transparency through autonomous policy explanation." In: *2017 12th ACM/IEEE International Conference on Human-Robot Interaction (HRI*. IEEE. 2017, pp. 303–312.

96. David Heckerman. "A tutorial on learning with Bayesian networks." In: *Innovations in Bayesian networks: Theory and applications* (2008), pp. 33–82.

97. Daniel Hein, Alexander Hentschel, Thomas Runkler, and Steffen Udluft. "Particle swarm optimization for generating interpretable fuzzy reinforcement learning policies." In: *Engineering Applications of Artificial Intelligence* 65 (2017), pp. 87–98.

98. Daniel Hein, Steffen Udluft, and Thomas A Runkler. "Interpretable policies for reinforcement learning by genetic programming." In: *Engineering Applications of Artificial Intelligence* 76 (2018), pp. 158–169.

99. Tue Herlau and Rasmus Larsen. "Reinforcement Learning of Causal Variables using Mediation Analysis." In: (2022).

100. Yvette Clarece Hester. *An analysis of the use and misuse of ANOVA*. Texas A&M University, 2000.

101. Johan Himberg, Aapo Hyvärinen, and Fabrizio Esposito. "Validating the independent components of neuroimaging time series via clustering and visualization." In: *Neuroimage* 22.3 (2004), pp. 1214–1222.

102. Sepp Hochreiter and Jürgen Schmidhuber. "Long short-term memory." In: *Neural computation* 9.8 (1997), pp. 1735–1780.

103. Robert R Hoffman, Shane T Mueller, Gary Klein, and Jordan Litman. "Metrics for explainable AI: Challenges and prospects." In: *arXiv preprint* arXiv:1812.04608 (2018).

104. Kurt Hornik, Maxwell Stinchcombe, and Halbert White. "Multilayer feedforward networks are universal approximators." In: *Neural networks* 2.5 (1989), pp. 359–366.

105. Jianfeng Huang, Plamen P Angelov, and Chengliang Yin. "Interpretable policies for reinforcement learning by empirical fuzzy sets." In: *Engineering Applications of Artificial Intelligence* 91 (2020), p. 103559.

106. Sandy H Huang, Kush Bhatia, Pieter Abbeel, and Anca D Dragan. "Establishing appropriate trust via critical states." In: *2018 IEEE/RSJ International Conference on Intelligent Robots and Systems (IROS)*. IEEE. 2018, pp. 3929–3936.

107. Tobias Huber, Dominik Schiller, and Elisabeth André. "Enhancing explainability of deep reinforcement learning through selective layer-wise relevance propagation." In: *Joint German/Austrian Conference on Artificial Intelligence (Künstliche Intelligenz)*. Springer. 2019, pp. 188–202.

108. Tobias Huber, Katharina Weitz, Elisabeth André, and Ofra Amir. "Local and global explanations of agent behavior: Integrating strategy summaries with saliency maps." In: *Artificial Intelligence* 301 (2021), p. 103571.

109. Aapo Hyvarinen. "Fast ICA for noisy data using Gaussian moments." In: *1999 IEEE international symposium on circuits and systems (ISCAS)*. Vol. 5. IEEE. 1999, pp. 57–61.

110. Aapo Hyvarinen, Juha Karhunen, and Erkki Oja. "Independent component analysis." In: *Studies in informatics and control* 11.2 (2002), pp. 205–207.

111. Aapo Hyvärinen, Kun Zhang, Shohei Shimizu, and Patrik O Hoyer. "Estimation of a structural vector autoregression model using non-gaussianity." In: *Journal of Machine Learning Research* 11.5 (2010).

112. Rahul Iyer, Yuezhang Li, Huao Li, Michael Lewis, Ramitha Sundar, and Katia Sycara. "Transparency and explanation in deep reinforcement learning neural networks." In: *Proceedings of the 2018 AAAI/ACM Conference on AI, Ethics, and Society*. 2018, pp. 144–150.

113. Robert A Jacobs, Michael I Jordan, Steven J Nowlan, and Geoffrey E Hinton. "Adaptive mixtures of local experts." In: *Neural computation* 3.1 (1991), pp. 79–87.

114. Alexis Jacq, Johan Ferret, Olivier Pietquin, and Matthieu Geist. "Lazy-mdps: Towards interpretable reinforcement learning by learning when to act." In: *arXiv preprint* arXiv:2203.08542 (2022).

115. Philip N Johnson-Laird. "Mental models and human reasoning." In: *Proceedings of the National Academy of Sciences* 107.43 (2010), pp. 18243–18250.

116. Anders Jonsson. "Deep reinforcement learning in medicine." In: *Kidney diseases* 5.1 (2019), pp. 18–22.

117. Ho-Taek Joo and Kyung-Joong Kim. "Visualization of deep reinforcement learning using grad-CAM: how AI plays atari games?" In: *2019 IEEE Conference on Games (CoG)*. IEEE. 2019, pp. 1–2.

118. Zoe Juozapaitis, Anurag Koul, Alan Fern, Martin Erwig, and Finale Doshi-Velez. "Explainable reinforcement learning via reward decomposition." In: *IJCAI/ECAI Workshop on explainable artificial intelligence*. 2019.

119. Daniel Kahneman. *Thinking, fast and slow*. macmillan, 2011.

120. Seyed Jalal Kazemitabar and Hamid Beigy. "Using strongly connected components as a basis for autonomous skill acquisition in reinforcement learning." In: *Advances in Neural Networks–ISNN 2009: 6th International Symposium on Neural Networks, ISNN 2009 Wuhan, China, May 26–29, 2009 Proceedings, Part I 6*. Springer. 2009, pp. 794–803.

121. James Kennedy and Russell Eberhart. "Particle swarm optimization." In: *Proceedings of ICNN'95-international conference on neural networks*. Vol. 4. IEEE. 1995, pp. 1942–1948.

122. Mohamed A Khamsi and William A Kirk. *An introduction to metric spaces and fixed point theory*. John Wiley & Sons, 2011.

123. Omar Khan, Pascal Poupart, and James Black. "Minimal sufficient explanations for factored markov decision processes." In: *Proceedings of the International Conference on Automated Planning and Scheduling*. Vol. 19. 2009, pp. 194–200.

124. Sangeet S Khemlani, Aron K Barbey, and Philip N Johnson-Laird. "Causal reasoning with mental models." In: *Frontiers in human neuroscience* 8 (2014), p. 849.

125. Mikko Kivelä, Alex Arenas, Marc Barthelemy, James P Gleeson, Yamir Moreno, and Mason A Porter. "Multilayer networks." In: *Journal of complex networks* 2.3 (2014), pp. 203–271.

126. Agneza Krajna, Mario Brcic, Tomislav Lipic, and Juraj Doncevic. "Explainability in reinforcement learning: perspective and position." In: *arXiv preprint* arXiv:2203.11547 (2022).

127. Werner Kritzinger, Matthias Karner, Georg Traar, Jan Henjes, and Wilfried Sihn. "Digital Twin in manufacturing: A categorical literature review and classification." In: *Ifac-PapersOnline* 51.11 (2018), pp. 1016–1022.

128. William H Kruskal and W Allen Wallis. "Use of ranks in one-criterion variance analysis." In: *Journal of the American statistical Association* 47.260 (1952), pp. 583–621.

129. Tejas D Kulkarni, Karthik Narasimhan, Ardavan Saeedi, and Josh Tenenbaum. "Hierarchical deep reinforcement learning: Integrating temporal abstraction and intrinsic motivation." In: *Advances in neural information processing systems* 29 (2016).

130. Solomon Kullback and Richard A Leibler. "On information and sufficiency." In: *The annals of mathematical statistics* 22.1 (1951), pp. 79–86.

131. Satyam Kumar, Mendhikar Vishal, and Vadlamani Ravi. "Explainable Reinforcement Learning on Financial Stock Trading using SHAP." In: *arXiv preprint* arXiv:2208.08790 (2022).

132. Joohyung Lee, Vladimir Lifschitz, and Fangkai Yang. "Action Language BC: Preliminary Report." In: *IJCAI*. Citeseer. 2013, pp. 983–989.

133. Henry Leung and Simon Haykin. "The complex backpropagation algorithm." In: *IEEE Transactions on signal processing* 39.9 (1991), pp. 2101–2104.

134. Howard Levene et al. "Contributions to probability and statistics." In: *Essays in honor of Harold Hotelling* 278 (1960), p. 292.

135. Rensis Likert. "A technique for the measurement of attitudes." In: *Archives of psychology* (1932).

136. Timothy P Lillicrap, Jonathan J Hunt, Alexander Pritzel, Nicolas Heess, Tom Erez, Yuval Tassa, David Silver, and Daan Wierstra. "Continuous control with deep reinforcement learning." In: *arXiv preprint* arXiv:1509.02971 (2015).

137. Zachary C Lipton. "The mythos of model interpretability: In machine learning, the concept of interpretability is both important and slippery." In: *Queue* 16.3 (2018), pp. 31–57.

138. Bing Liu, Yiyuan Xia, and Philip S Yu. "Clustering through decision tree construction." In: *Proceedings of the ninth international conference on Information and knowledge management*. 2000, pp. 20–29.

139. Guiliang Liu, Oliver Schulte, Wang Zhu, and Qingcan Li. "Toward interpretable deep reinforcement learning with linear model u-trees." In: *Joint European Conference on Machine Learning and Knowledge Discovery in Databases*. Springer. 2018, pp. 414–429.

140. Xiao-Yang Liu, Zihan Ding, Sem Borst, and Anwar Walid. "Deep reinforcement learning for intelligent transportation systems." In: *arXiv preprint* arXiv:1812.00979 (2018).

141. Yongkui Liu, He Xu, Ding Liu, and Lihui Wang. "A digital twin-based sim-to-real transfer for deep reinforcement learning-enabled industrial robot grasping." In: *Robotics and Computer-Integrated Manufacturing* 78 (2022), p. 102365.

142. Stuart Lloyd. "Least squares quantization in PCM." In: *IEEE transactions on information theory* 28.2 (1982), pp. 129–137.

143. Wenhao Lu, Xufeng Zhao, Thilo Fryen, Jae Hee Lee, Mengdi Li, Sven Magg, and Stefan Wermter. "Causal State Distillation for Explainable Reinforcement Learning." In: *Causal Learning and Reasoning*. PMLR. 2024, pp. 106–142.

144. Scott M Lundberg and Su-In Lee. "A unified approach to interpreting model predictions." In: *Advances in neural information processing systems* 30 (2017).

145. Daoming Lyu, Fangkai Yang, Bo Liu, and Steven Gustafson. "SDRL: interpretable and data-efficient deep reinforcement learning leveraging symbolic planning." In: *Proceedings of the AAAI Conference on Artificial Intelligence*. Vol. 33. 01. 2019, pp. 2970–2977.

146. Zhihao Ma, Yuzheng Zhuang, Paul Weng, Hankz Hankui Zhuo, Dong Li, Wulong Liu, and Jianye Hao. "Learning Symbolic Rules for Interpretable Deep Reinforcement Learning." In: *arXiv preprint* arXiv:2103.08228 (2021).

147. Laurens Van der Maaten and Geoffrey Hinton. "Visualizing data using t-SNE." In: *Journal of machine learning research* 9.11 (2008).

148. Prashan Madumal, Tim Miller, Liz Sonenberg, and Frank Vetere. "Distal explanations for explainable reinforcement learning agents." In: *arXiv preprint* arXiv:2001.10284 (2020).

149. Prashan Madumal, Tim Miller, Liz Sonenberg, and Frank Vetere. "Explainable reinforcement learning through a causal lens." In: *Proceedings of the AAAI conference on artificial intelligence*. Vol. 34. 03. 2020, pp. 2493–2500.

150. Shie Mannor, Ishai Menache, Amit Hoze, and Uri Klein. "Dynamic abstraction in reinforcement learning via clustering." In: *Proceedings of the twenty-first international conference on Machine learning*. 2004, p. 71.

151. Joe McCalmon, Thai Le, Sarra Alqahtani, and Dongwon Lee. "CAPS: Comprehensible Abstract Policy Summaries for Explaining Reinforcement Learning Agents." In: *Proceedings of the 21st International Conference on Autonomous Agents and Multiagent Systems*. 2022, pp. 889–897.

152. James McCarthy, Rahul Nair, Elizabeth Daly, Radu Marinescu, and Ivana Dusparic. "Boolean Decision Rules for Reinforcement Learning Policy Summarisation." In: *arXiv preprint* arXiv:2207.08651 (2022).

153. Larry Medsker and Lakhmi C Jain. *Recurrent neural networks: design and applications*. CRC press, 1999.

154. Ishai Menache, Shie Mannor, and Nahum Shimkin. "Q-cut—dynamic discovery of subgoals in reinforcement learning." In: *Machine Learning: ECML 2002: 13th European*

*Conference on Machine Learning Helsinki, Finland, August 19–23, 2002 Proceedings 13*. Springer. 2002, pp. 295–306.

155. Matheus RF Mendonca, Artur Ziviani, and André MS Barreto. "Graph-based skill acquisition for reinforcement learning." In: *ACM Computing Surveys (CSUR)* 52.1 (2019), pp. 1–26.

156. Hugo Mercier and Dan Sperber. "Why do humans reason? Arguments for an argumentative theory." In: *Behavioral and brain sciences* 34.2 (2011), pp. 57–74.

157. Rudy Milani. "Towards an Automatic Ensemble Methodology for Explainable Reinforcement Learning." *2024 IEEE 14th Annual Computing and Communication Workshop and Conference, CCWC 2024*, (2024), pp. 301–307. https://doi.org/10.1109/CCWC60891.2024.10427957.

158. Rudy Milani, Maximilian Moll, and Renato De Leone. "Detection of Important States through an Iterative Q-value Algorithm for Explainable Reinforcement Learning." In: *Proceedings of the Annual Hawaii International Conference on System Sciences*, (2024), pp. 1401–1408.

159. Rudy Milani, Maximilian Moll, Renato De Leone, and Stefan Pickl. "A Bayesian Network Approach to Explainable Reinforcement Learning with Distal Information." *Sensors*, vol. 23, no. 4, p. 2013, (2023), https://doi.org/10.3390/s23042013.

160. Rudy Milani, Maximilian Moll, Renato De Leone, and Stefan Pickl. "A Multigraph and Multiplex Approach for the Recognition of Important States in MDPs." In: (submitted).

161. Rudy Milani, Maximilian Moll, and Stefan Pickl. "Advances in Explainable Reinforcement Learning: An Intelligent Transportation Systems Perspective." In: *Explainable Artificial Intelligence for Intelligent Transportation Systems*. CRC press, pp. 93—118, Sep. 2023, https://doi.org/10.1201/9781003324140-5.

162. Rudy Milani, Marian Sorin Nistor, Maximilian Moll, and Stefan Pickl. "On the Correlation and Predictability of Topological Measures in Transportation Networks." In: *Operations Research Forum*. Vol. 6. 2. Springer. 2025, p. 82. https://doi.org/10.1007/s43069-025-00471-8.

163. Stephanie Milani, Zhicheng Zhang, Nicholay Topin, Zheyuan Ryan Shi, Charles Kamhoua, Evangelos E Papalexakis, and Fei Fang. "Interpretable Multi-Agent Reinforcement Learning with Decision-Tree Policies." In: *Explainable Agency in Artificial Intelligence*. CRC Press, 2024, pp. 86–120.

164. Tim Miller. "Explanation in artificial intelligence: Insights from the social sciences." In: *Artificial intelligence* 267 (2019), pp. 1–38.

165. Indrajeet Mishra, Giang Dao, and Minwoo Lee. "Visual sparse Bayesian reinforcement learning: a framework for interpreting what an agent has learned." In: *2018 IEEE Symposium Series on Computational Intelligence (SSCI)*. IEEE. 2018, pp. 1427–1434.

166. Volodymyr Mnih, Koray Kavukcuoglu, David Silver, Alex Graves, Ioannis Antonoglou, Daan Wierstra, and Martin Riedmiller. "Playing atari with deep reinforcement learning." In: *arXiv preprint* arXiv:1312.5602 (2013).

167. Volodymyr Mnih, Koray Kavukcuoglu, David Silver, Andrei A Rusu, Joel Veness, Marc G Bellemare, Alex Graves, Martin Riedmiller, Andreas K Fidjeland, Georg Ostrovski, et al. "Human-level control through deep reinforcement learning." In: *nature* 518.7540 (2015), pp. 529–533.

168. C Molnar. *Interpretable Machine Learning: A Guide for Making Black Box Models Explainable. Published online*. 2019.

169. Grégoire Montavon, Wojciech Samek, and Klaus-Robert Müller. "Methods for interpreting and understanding deep neural networks." In: *Digital signal processing* 73 (2018), pp. 1–15.

170. Andrew William Moore. *Efficient memory-based learning for robot control*. Tech. rep. University of Cambridge, Computer Laboratory, 1990.

171. Parham Moradi, Mohammad Ebrahim Shiri, and Negin Entezari. "Automatic skill acquisition in reinforcement learning agents using connection bridge centrality." In: *Communication and Networking: International Conference, FGCN 2010, Held as Part of the Future Generation Information Technology Conference, FGIT 2010, Jeju Island, Korea, December 13–15, 2010. Proceedings, Part II*. Springer. 2010, pp. 51–62.

172. Zai Mueller-Zhang, Pablo Oliveira Antonino, and Thomas Kuhn. "Integrated planning and scheduling for customized production using digital twins and reinforcement learning." In: *IFAC-PapersOnLine* 54.1 (2021), pp. 408–413.

173. Bonnie M Muir. "Trust between humans and machines, and the design of decision aids." In: *International journal of man-machine studies* 27.5–6 (1987), pp. 527–539.

174. Bonnie M Muir. "Trust in automation: Part I. Theoretical issues in the study of trust and human intervention in automated systems." In: *Ergonomics* 37.11 (1994), pp. 1905–1922.

175. Prakash M Nadkarni, Lucila Ohno-Machado, and Wendy W Chapman. "Natural language processing: an introduction." In: *Journal of the American Medical Informatics Association* 18.5 (2011), pp. 544–551.

176. AS Nemirovsky. "Optimization II. Numerical methods for nonlinear continuous optimization." In: (1999).

177. Mark EJ Newman. "Analysis of weighted networks." In: *Physical review E* 70.5 (2004), p. 056131.

178. Tien T Nguyen, Pik-Mai Hui, F Maxwell Harper, Loren Terveen, and Joseph A Konstan. "Exploring the filter bubble: the effect of using recommender systems on content diversity." In: *Proceedings of the 23rd international conference on World wide web*. 2014, pp. 677–686.

179. Roxana Pamfil, Nisara Sriwattanaworachai, Shaan Desai, Philip Pilgerstorfer, Konstantinos Georgatzis, Paul Beaumont, and Bryon Aragam. "Dynotears: Structure learning from time-series data." In: *International Conference on Artificial Intelligence and Statistics*. Pmlr. 2020, pp. 1595–1605.

180. Nick Pawlowski, Daniel Coelho de Castro, and Ben Glocker. "Deep structural causal models for tractable counterfactual inference." In: *Advances in Neural Information Processing Systems* 33 (2020), pp. 857–869.

181. Judea Pearl. "Causal diagrams for empirical research." In: *Biometrika* 82.4 (1995), pp. 669–688.

182. Judea Pearl. "Causation, action, and counterfactuals." In: *Logic and Scientific Methods: Volume One of the Tenth International Congress of Logic, Methodology and Philosophy of Science, Florence, August 1995*. Springer. 1997, pp. 355–375.

183. Judea Pearl. "Graphs, causality, and structural equation models." In: *Sociological Methods & Research* 27.2 (1998), pp. 226–284.

184. Judea Pearl. *Causality*. Cambridge university press, 2009.

185. Judea Pearl. "An introduction to causal inference." In: *The international journal of biostatistics* 6.2 (2010).

186. Judea Pearl. "Graphical models, causality, and intervention." In: (2011).

187. Judea Pearl. "The causal mediation formula—a guide to the assessment of pathways and mechanisms." In: *Prevention science* 13.4 (2012), pp. 426–436.

188. José M Pena, Jose Antonio Lozano, and Pedro Larranaga. "An empirical comparison of four initialization methods for the k-means algorithm." In: Pattern recognition letters 20.10 (1999), pp. 1027–1040.

189. Oskar Perron. "Mathematische Annalen." In: *Zur Theorie der Matrices* 64.2 (1907), pp. 248–263.

190. Jonas Peters, Dominik Janzing, and Bernhard Schölkopf. *Elements of causal inference: foundations and learning algorithms.* The MIT Press, 2017.

191. Rakesh Kumar Phanden, Priavrat Sharma, and Anubhav Dubey. "A review on simulation in digital twin for aerospace, manufacturing and robotics." In: *Materials today: proceedings* 38 (2021), pp. 174–178.

192. S Unnikrishna Pillai, Torsten Suel, and Seunghun Cha. "The Perron-Frobenius theorem: some of its applications." In: *IEEE Signal Processing Magazine* 22.2 (2005), pp. 62–75.

193. Erika Puiutta and Eric Veith. "Explainable reinforcement learning: A survey." In: (2020), pp. 77–95.

194. David V Pynadath, Michael J Barnes, Ning Wang, and Jessie YC Chen. "Transparency communication for machine learning in human-automation interaction." In: *Human and machine learning.* Springer, 2018, pp. 75–90.

195. Ali Ajdari Rad, Martin Hasler, and Parham Moradi. "Automatic skill acquisition in Reinforcement Learning using connection graph stability centrality." In: *Proceedings of 2010 IEEE International Symposium on Circuits and Systems.* IEEE. 2010, pp. 697–700.

196. Gabriëlle Ras, Ning Xie, and Derek Doran. *Explainable Deep Learning: A Field Guide for the Uninitiated.* 2022, pp. 329–396.

197. Marco Tulio Ribeiro, Sameer Singh, and Carlos Guestrin. ""Why should i trust you?" Explaining the predictions of any classifier." In: *Proceedings of the 22nd ACM SIGKDD international conference on knowledge discovery and data mining.* 2016, pp. 1135–1144.

198. Marco Tulio Ribeiro, Sameer Singh, and Carlos Guestrin. "Model-agnostic interpretability of machine learning." In: *arXiv preprint* arXiv:1606.05386 (2016).

199. Finn Rietz, Sven Magg, Fredrik Heintz, Todor Stoyanov, Stefan Wermter, and Johannes A Stork. "Hierarchical goals contextualize local reward decomposition explanations." In: *Neural Computing and Applications* (2022), pp. 1–12.

200. Stefano Giovanni Rizzo, Giovanna Vantini, and Sanjay Chawla. "Reinforcement learning with explainability for traffic signal control." In: *2019 IEEE Intelligent Transportation Systems Conference (ITSC).* IEEE. 2019, pp. 3567–3572.

201. Peter J Rousseeuw. "Silhouettes: a graphical aid to the interpretation and validation of cluster analysis." In: *Journal of computational and applied mathematics* 20 (1987), pp. 53–65.

202. Donald B Rubin. "Estimating causal effects of treatments in randomized and nonrandomized studies." In: *Journal of educational Psychology* 66.5 (1974), p. 688.

203. Cynthia Rudin. "Stop explaining black box machine learning models for high stakes decisions and use interpretable models instead." In: *Nature Machine Intelligence* 1 (5 May 2019), pp. 206–215. ISSN: 25225839. https://doi.org/10.1038/s42256-019-0048-x.

204. Gavin A Rummery and Mahesan Niranjan. *On-line Q-learning using connectionist systems*. Vol. 37. University of Cambridge, Department of Engineering Cambridge, UK, 1994.

205. Gavin Adrian Rummery. "Problem solving with reinforcement learning." PhD thesis. Citeseer, 1995.

206. Stuart J Russell and Andrew Zimdars. "Q-decomposition for reinforcement learning agents." In: *Proceedings of the 20th International Conference on Machine Learning (ICML-03)*. 2003, pp. 656–663.

207. Gert Sabidussi. "The centrality index of a graph." In: *Psychometrika* 31.4 (1966), pp. 581–603.

208. Aatmaj Amol Salunke. "Reinforcement learning empowered digital twins: Pioneering smart cities towards optimal urban dynamics." In: *EPRA International Journal of Research & Development (IJRD)* 10 (2023).

209. Wojciech Samek and Klaus Robert Müller. *Towards Explainable Artificial Intelligence*. 2019. https://doi.org/10.1007/978-3-030-28954-6_1.

210. Daniel Schunk and Cornelia Betsch. "Explaining heterogeneity in utility functions by individual differences in decision modes." In: *Journal of Economic Psychology* 27.3 (2006), pp. 386–401.

211. Ramprasaath R Selvaraju, Michael Cogswell, Abhishek Das, Ramakrishna Vedantam, Devi Parikh, and Dhruv Batra. "Grad-cam: Visual explanations from deep networks via gradient-based localization." In: *Proceedings of the IEEE international conference on computer vision*. 2017, pp. 618–626.

212. Pedro Sequeira and Melinda Gervasio. "Interestingness elements for explainable reinforcement learning: Understanding agents' capabilities and limitations." In: *Artificial Intelligence* 288 (2020), p. 103367.

213. Samuel Sanford Shapiro and Martin B Wilk. "An analysis of variance test for normality (complete samples)." In: *Biometrika* 52.3–4 (1965), pp. 591–611.

214. Jing Shen, Guochang Gu, and Haibo Liu. "Automatic option generation in hierarchical reinforcement learning via immune clustering." In: *2006 1st International Symposium on Systems and Control in Aerospace and Astronautics*. IEEE. 2006, 4–pp.

215. Shohei Shimizu, Patrik O Hoyer, Aapo Hyvärinen, Antti Kerminen, and Michael Jordan. "A linear non-Gaussian acyclic model for causal discovery." In: *Journal of Machine Learning Research* 7.10 (2006).

216. Shohei Shimizu, Takanori Inazumi, Yasuhiro Sogawa, Aapo Hyvarinen, Yoshinobu Kawahara, Takashi Washio, Patrik O Hoyer, Kenneth Bollen, and Patrik Hoyer. "DirectLiNGAM: A direct method for learning a linear non-Gaussian structural equation model." In: *Journal of Machine Learning Research-JMLR* 12.Apr (2011), pp. 1225–1248.

217. Farzaneh Shoeleh and Masoud Asadpour. "Graph based skill acquisition and transfer learning for continuous reinforcement learning domains." In: *Pattern Recognition Letters* 87 (2017), pp. 104–116.

218. Farzaneh Shoeleh and Masoud Asadpour. "Skill based transfer learning with domain adaptation for continuous reinforcement learning domains." In: *Applied Intelligence* 50 (2020), pp. 502–518.

219. Tianmin Shu, Caiming Xiong, and Richard Socher. "Hierarchical and interpretable skill acquisition in multi-task reinforcement learning." In: *arXiv preprint* arXiv:1712.07294 (2017).

220. Andrew Silva, Matthew Gombolay, Taylor Killian, Ivan Jimenez, and Sung-Hyun Son. "Optimization methods for interpretable differentiable decision trees applied to reinforcement learning." In: *International conference on artificial intelligence and statistics*. PMLR. 2020, pp. 1855–1865.

221. Ozgur Simsek. *Behavioral building blocks for autonomous agents: description, identification, and learning*. University of Massachusetts Amherst, 2008.

222. Özgür Şimşek and Andrew G Barto. "Using relative novelty to identify useful temporal abstractions in reinforcement learning." In: *Proceedings of the twenty-first international conference on Machine learning*. 2004, p. 95.

223. Özgür Şimşek and Andrew G Barto. "Betweenness centrality as a basis for forming skills." In: (2007).

224. Özgür Şimşek and Andrew Barto. "Skill characterization based on betweenness." In: *Advances in neural information processing systems* 21 (2008).

225. Özgür Şimşek, Alicia P Wolfe, and Andrew G Barto. "Identifying useful subgoals in reinforcement learning by local graph partitioning." In: *Proceedings of the 22nd international conference on Machine learning*. 2005, pp. 816–823.

226. Luis Solá, Miguel Romance, Regino Criado, Julio Flores, Alejandro García del Amo, and Stefano Boccaletti. "Eigenvector centrality of nodes in multiplex networks." In: *Chaos: An Interdisciplinary Journal of Nonlinear Science* 23.3 (2013).

227. David J Spiegelhalter and Steffen L Lauritzen. "Sequential updating of conditional probabilities on directed graphical structures." In: *Networks* 20.5 (1990), pp. 579–605.

228. Peter Spirtes, Clark Glymour, and Richard Scheines. *Causation, prediction, and search*. MIT press, 2001.

229. Alberto Suárez and James F Lutsko. "Globally optimal fuzzy decision trees for classification and regression." In: *IEEE Transactions on Pattern Analysis and Machine Intelligence* 21.12 (1999), pp. 1297–1311.

230. Mukund Sundararajan, Ankur Taly, and Qiqi Yan. "Axiomatic attribution for deep networks." In: *International conference on machine learning*. PMLR. 2017, pp. 3319–3328.

231. Richard S Sutton and Andrew G Barto. *Reinforcement learning: An introduction*. MIT press, 2018.

232. Richard S Sutton, Doina Precup, and Satinder Singh. "Between MDPs and semi-MDPs: A framework for temporal abstraction in reinforcement learning." In: *Artificial intelligence* 112.1–2 (1999), pp. 181–211.

233. Csaba Szepesvári. "The asymptotic convergence-rate of Q-learning." In: *Advances in neural information processing systems* 10 (1997).

234. Aaquib Tabrez, Shivendra Agrawal, and Bradley Hayes. "Explanation-based reward coaching to improve human performance via reinforcement learning." In: *2019 14th ACM/IEEE International Conference on Human-Robot Interaction (HRI)*. IEEE. 2019, pp. 249–257.

235. Nasrin Taghizadeh and Hamid Beigy. "A novel graphical approach to automatic abstraction in reinforcement learning." In: *Robotics and Autonomous Systems* 61.8 (2013), pp. 821–835.

236. Yoshiki Takagi, Roderick Tabalba, Nurit Kirshenbaum, and Jason Leigh. "Abstracted Trajectory Visualization for Explainability in Reinforcement Learning." In: *arXiv preprint* arXiv:2402.07928 (2024).

237. Michael E Tipping. "Sparse Bayesian learning and the relevance vector machine." In: *Journal of machine learning research* 1.Jun (2001), pp. 211–244.

238. Erico Tjoa and Cuntai Guan. "A Survey on Explainable Artificial Intelligence (XAI): Toward Medical XAI." In: *IEEE Transactions on Neural Networks and Learning Systems* 32 (11 Nov. 2021), pp. 4793–4813. ISSN: 21622388. https://doi.org/10.1109/TNNLS.2020.3027314.

239. Lisa Torrey and Matthew Taylor. "Teaching on a budget: Agents advising agents in reinforcement learning." In: *Proceedings of the 2013 international conference on Autonomous agents and multi-agent systems*. 2013, pp. 1053–1060.

240. John Tsitsiklis and Benjamin Van Roy. "Analysis of temporal-diffference learning with function approximation." In: *Advances in neural information processing systems* 9 (1996).

241. Francesco Tudisco, Francesca Arrigo, and Antoine Gautier. "Node and layer eigenvector centralities for multiplex networks." In: *SIAM Journal on Applied Mathematics* 78.2 (2018), pp. 853–876.

242. John W Tukey. "Comparing individual means in the analysis of variance." In: *Biometrics* (1949), pp. 99–114.

243. William TB Uther and Manuela M Veloso. "Tree based discretization for continuous state space reinforcement learning." In: *Aaai/iaai* 98 (1998), pp. 769–774.

244. Peter Vamplew, Richard Dazeley, Cameron Foale, Sally Firmin, and Jane Mummery. "Human-aligned artificial intelligence is a multiobjective problem." In: *Ethics and Information Technology* 20.1 (2018), pp. 27–40.

245. Marko Vasić, Andrija Petrović, Kaiyuan Wang, Mladen Nikolić, Rishabh Singh, and Sarfraz Khurshid. "MoËT: Mixture of Expert Trees and its application to verifiable reinforcement learning." In: *Neural Networks* 151 (2022), pp. 34–47.

246. Ashish Vaswani, Noam Shazeer, Niki Parmar, Jakob Uszkoreit, Llion Jones, Aidan N Gomez, Lukasz Kaiser, and Illia Polosukhin. "Attention is all you need." In: Advances in *neural information processing systems* 30 (2017).

247. Abhinav Verma, Vijayaraghavan Murali, Rishabh Singh, Pushmeet Kohli, and Swarat Chaudhuri. "Programmatically interpretable reinforcement learning." In: *International Conference on Machine Learning*. PMLR. 2018, pp. 5045–5054.

248. Mathurin Videau, Alessandro Leite, Olivier Teytaud, and Marc Schoenauer. "Multi-objective genetic programming for explainable reinforcement learning." In: *European Conference on Genetic Programming (Part of EvoStar)*. Springer. 2022, pp. 278–293.

249. Kristel Vignery and Wim Laurier. "A methodology and theoretical taxonomy for centrality measures: What are the best centrality indicators for student networks?" In: *Plos one* 15.12 (2020), e0244377.

250. Sergei Volodin. "Causeoccam: Learning interpretable abstract representations in reinforcement learning environments via model sparsity." In: (2021).

251. Kevin L Voogd, Jean Pierre Allamaa, Javier Alonso-Mora, and Tong Duy Son. "Reinforcement learning from simulation to real world autonomous driving using digital twin." In: *IFAC-PapersOnLine* 56.2 (2023), pp. 1510–1515.

252. George A. Vouros. "Explainable Deep Reinforcement Learning: State of the Art and Challenges." In: *ACM Computing Surveys* (Mar. 2022). ISSN: 0360-0300. https://doi.org/10.1145/3527448.

253. Jasper van der Waa, Jurriaan van Diggelen, Karel van den Bosch, and Mark Neerincx. "Contrastive explanations for reinforcement learning in terms of expected consequences." In: *arXiv preprint* arXiv:1807.08706 (2018).

254. Junpeng Wang, Liang Gou, Han-Wei Shen, and Hao Yang. "Dqnviz: A visual analytics approach to understand deep q-networks." In: *IEEE transactions on visualization and computer graphics* 25.1 (2018), pp. 288–298.

255. Xiaoxiao Wang, Fanyu Meng, Xin Liu, Zhaodan Kong, and Xin Chen. "Causal explanation for reinforcement learning: Quantifying state and temporal importance." In: *Applied Intelligence* 53.19 (2023), pp. 22546–22564.

256. Yuyao Wang, Masayoshi Mase, and Masashi Egi. "Attribution-based Salience Method towards Interpretable Reinforcement Learning." In: *AAAI Spring Symposium: Combining Machine Learning with Knowledge Engineering (1)*. 2020.

257. Zhikang T Wang and Masahito Ueda. "Convergent and efficient deep Q network algorithm." In: *arXiv preprint* arXiv:2106.15419 (2021).

258. Christopher JCH Watkins and Peter Dayan. "Q-learning." In: *Machine learning* 8 (1992), pp. 279–292.

259. Bernard Lewis Welch. "On the comparison of several mean values: an alternative approach." In: *Biometrika* 38.3/4 (1951), pp. 330–336.

260. Adrian Weller. "Transparency: motivations and challenges." In: *Explainable AI: interpreting, explaining and visualizing deep learning*. Springer, 2019, pp. 23–40.

261. Paul J Werbos. "Backpropagation through time: what it does and how to do it." In: *Proceedings of the IEEE* 78.10 (1990), pp. 1550–1560.

262. Marco A Wiering and Martijn Van Otterlo. "Reinforcement learning." In: *Adaptation, learning, and optimization* 12.3 (2012), p. 729.

263. Ronald J Williams. "Simple statistical gradient-following algorithms for connectionist reinforcement learning." In: *Machine learning* 8 (1992), pp. 229–256.

264. Sewall Wright. "Correlation and causation." In: *Journal of agricultural research* 20.7 (1921), p. 557.

265. Kaishu Xia, Christopher Sacco, Max Kirkpatrick, Clint Saidy, Lam Nguyen, Anil Kircaliali, and Ramy Harik. "A digital twin to train deep reinforcement learning agent for smart manufacturing plants: Environment, interfaces and intelligence." In: *Journal of Manufacturing Systems* 58 (2021), pp. 210–230.

266. Zhao Yang, Song Bai, Li Zhang, and Philip HS Torr. "Learn to interpret atari agents." In: *arXiv preprint* arXiv:1812.11276 (2018).

267. Zhongwei Yu, Jingqing Ruan, and Dengpeng Xing. "Explainable reinforcement learning via a causal world model." In: *arXiv preprint* arXiv:2305.02749 (2023).

268. Hao Yuan, Haiyang Yu, Shurui Gui, and Shuiwang Ji. "Explainability in graph neural networks: A taxonomic survey." In: *arXiv preprint* arXiv:2012.15445 (2020).

269. Vinicius Zambaldi, David Raposo, Adam Santoro, Victor Bapst, Yujia Li, Igor Babuschkin, Karl Tuyls, David Reichert, Timothy Lillicrap, Edward Lockhart, et al. "Deep reinforcement learning with relational inductive biases." In: *International conference on learning representations*. 2018.

270. Zhongheng Zhang et al. "Reinforcement learning in clinical medicine: a method to optimize dynamic treatment regime over time." In: *Annals of translational medicine* 7.14 (2019).

271. Min Zheng, Jessecae K Marsh, Jeffrey V Nickerson, and Samantha Kleinberg. "How causal information affects decisions." In: *Cognitive Research: Principles and Implications* 5 (2020), pp. 1–24.

272. Xun Zheng, Chen Dan, Bryon Aragam, Pradeep Ravikumar, and Eric Xing. "Learning sparse nonparametric dags." In: *International Conference on Artificial Intelligence and Statistics*. Pmlr. 2020, pp. 3414–3425.

273. Jinming Zou, Yi Han, and Sung-Sau So. "Overview of artificial neural networks." In: *Artificial neural networks: methods and applications* (2009), pp. 14–22.

274. Michael J Zyphur and Frederick L Oswald. "Bayesian estimation and inference: A user's guide." In: *Journal of Management* 41.2 (2015), pp. 390–420.